Languages in Space and Time

This cross-disciplinary volume provides an overview of how complexity theory and the tools of statistical mechanics can be applied to linguistic problems to help reveal language groups, and to model the evolution and competition of languages in space and time. On the other hand, it aims to provide some new knowledge of linguistics to mathematicians/physicists. It demonstrates the complexity of linguistic databases and provides a mathematical toolkit for analyzing and extracting useful information from them. Various case studies show how mathematical analysis and modeling help conceptualize empirical facts better than what can be learned through a mere ethnographic view, while the interpretation of facts through sociolinguistics may promote mathematical modeling that predicts and represents reality more accurately. The book should contribute to building a bridge between linguists and mathematical modelers interested in linguistics.

Marco Patriarca is Senior Researcher at the National Institute of Chemical Physics and Biophysics, Tallinn, Estonia. His areas of interest include stochastic processes, diffusion processes, Brownian motion, condensed matter, quantum mechanics, and physics of language.

Els Heinsalu is Senior Researcher at the National Institute of Chemical Physics and Biophysics, Tallinn, Estonia. Her areas of interest are statistical physics and its applications to complex systems, modeling of language competition, and stochastic processes.

Jean Léo Léonard is Professor at the University of Montpellier–3 (Paul-Valéry). A linguist with wide experience in many languages, he has worked in Uralic linguistics, Basque, Romance dialectology, and various Central-American languages.

Physics of Society: Econophysics and Sociophysics

This book series is aimed at introducing readers to the recent developments in physics inspired modeling of economic and social systems. Socio-economic systems are increasingly being identified as 'interacting many-body dynamical systems' very much similar to the physical systems, studied over several centuries now. Econophysics and sociophysics, as interdisciplinary subjects, view the dynamics of markets and society in general as those of physical systems. This will be a series of books written by eminent academicians, researchers and subject experts in the field of physics, mathematics, finance, sociology, management, and economics.

This new series brings out research monographs and course books useful for the students and researchers across disciplines, both from physical and social science disciplines, including economics.

Series Editors:

Bikas K. Chakrabarti
Emeritus Professor, Saha Institute of
Nuclear Physics, Kolkata, India
Visiting Professor of Economics, Indian
Statistical Institute, Kolkata, India

Mauro Gallegati
Professor of Economics
Polytechnic University of Marche
Italy

Alan Kirman
Professor Emeritus of Economics
University of AixMarseille III, Marseille
France

H. Eugene Stanley
William Fairfield Warren Distinguished Professor
Boston University, Boston
USA

Editorial Board Members:

Frédéric Abergel
Centrale Supélec
Université Paris-Saclay,
Gif-Sur-Yvette
France

Hideaki Aoyama
Professor of Physics
Kyoto University, Kyoto
Japan

Anirban Chakraborti
Professor of Physics
School of Computational and Integrative Sciences
Jawaharlal Nehru University, New Delhi
India

Satya Ranjan Chakravarty
Professor of Economics
Indian Statistical Institute, Kolkata
India

Arnab Chatterjee
TCS Innovation Labs, Delhi
India

Shu-Heng Chen
Professor of Economics and Computer Science
Director, AIECON Research Center
National Chengchi University, Taipei
Taiwan

Domenico Delli Gatti
Professor of Economics
Catholic University, Milan
Italy

Kausik Gangopadhyay
Professor of Economics
Indian Institute of Management, Kozhikode
India

Cars Hommes
Professor of Economics
Amsterdam School of Economics
University of Amsterdam
Director, Center for Nonlinear Dynamics in
Economics and Finance (CeNDEF), Amsterdam
Netherlands

Giulia Iori
Professor of Economics
School of Social Science
City University, London
United Kingdom

Teisei Kaizoji
Professor of Economics
Department of Economics and Business
International Christian University, Tokyo
Japan

Kimmo Kaski
Professor of Physics
Dean, School of Science
Aalto University, Espoo
Finland

János Kertész
Professor of Physics
Center for Network Science
Central European University, Budapest
Hungary

Akira Namatame
Professor of Computer Science and Economics
Department of Computer Science
National Defense Academy, Yokosuka
Japan

Parongama Sen
Professor of Physics
University of Calcutta, Kolkata
India

Sitabhra Sinha
Professor of Physics
Institute of Mathematical Science, Chennai
India

Victor Yakovenko
Professor of Physics
University of Maryland, College Park
USA

Physics of Society: Published Titles

- *Interactive Macroeconomics: Stochastic Aggregate Dynamics with Heterogeneous and Interacting Agents* by Corrado Di Guilmi, Simone Landini, and Mauro Gallegati
- *Limit Order Books* by Frédéric Abergel, Marouane Anane, Anirban Chakraborti, Aymen Jedidi, and Ioane Muni Toke
- *Macro-Econophysics: New Studies on Economic Networks and Synchronization* by Hideaki Aoyama, Yoshi Fujiwara, Yuichi Ikeda, Hiroshi Iyetomi, Wataru Souma, and Hiroshi Yoshikawa

Languages in Space and Time

Models and Methods from Complex Systems Theory

Marco Patriarca

Els Heinsalu

Jean Léo Léonard

CAMBRIDGE
UNIVERSITY PRESS

University Printing House, Cambridge CB2 8BS, United Kingdom

One Liberty Plaza, 20th Floor, New York, NY 10006, USA

477 Williamstown Road, Port Melbourne, VIC 3207, Australia

314 to 321, 3rd Floor, Plot No.3, Splendor Forum, Jasola District Centre, New Delhi 110025, India

79 Anson Road, #06–04/06, Singapore 079906

Cambridge University Press is part of the University of Cambridge.

It furthers the University's mission by disseminating knowledge in the pursuit of education, learning and research at the highest international levels of excellence.

www.cambridge.org
Information on this title: www.cambridge.org/9781108480659

© Cambridge University Press 2020

This publication is in copyright. Subject to statutory exception and to the provisions of relevant collective licensing agreements, no reproduction of any part may take place without the written permission of Cambridge University Press.

First published 2020

Printed in India by Nutech Print Services, New Delhi 110020

A catalogue record for this publication is available from the British Library

Library of Congress Cataloging-in-Publication Data

Names: Patriarca, Marco, author. | Heinsalu, Els, author. | Léonard, Jean Léo, author.
Title: Languages in space and time : models and methods from complex systems theory / Marco Patriarca, Els Heinsalu, Jean Léo Léonard.
Description: First edition. | New York : Cambridge University Press, 2020. | Series: Physics of society: econophysics and sociophysics | Includes bibliographical references and index.
Identifiers: LCCN 2019058617 (print) | LCCN 2019058618 (ebook) | ISBN 9781108480659 (hardback) | ISBN 9781108480659 (pdf)
Subjects: LCSH: Mathematical linguistics. | Computational linguistics.
Classification: LCC P138 .P38 2020 (print) | LCC P138 (ebook) | DDC 410.1/51–dc23 LC record available at https://lccn.loc.gov/2019058617
LC ebook record available at https://lccn.loc.gov/2019058618

ISBN 978-1-108-48065-9 Hardback

Cambridge University Press has no responsibility for the persistence or accuracy of URLs for external or third-party internet websites referred to in this publication, and does not guarantee that any content on such websites is, or will remain, accurate or appropriate.

Contents

List of Figures — xi
List of Tables — xv
Preface — xvii

1 Introduction — 1
  1.1 Introduction to Complexity — 1
  1.2 Aim of the Book — 4
  1.3 The Readership — 5
  1.4 Structure of the Book — 6

PART ONE: REVEALING LANGUAGE GROUPS

2 Language and Languages — 9
  2.1 Language Evolution — 11
  2.2 Language Change — 13
  2.3 Mechanisms of Sound Change — 16
  2.4 Complex Networks — 21
    2.4.1 General considerations about complex networks — 21
    2.4.2 Features of a complex network — 23
    2.4.3 Types of complex networks — 24
  2.5 Trees vs. Networks — 28
    2.5.1 Epistemological Consequences: The Open Taxonomy Challenge — 34

3 Comparison Based on String Metric — 41
  3.1 Levenshtein Distance — 41
  3.2 Normalized Levenshtein Distance — 42
  3.3 Levenshtein Distance for Language Comparison — 43
  3.4 Similarity Index — 44
  3.5 Case Study: The Mazatec Language — 45
    3.5.1 Mazatec: Introduction — 46
    3.5.2 Mazatec: A short overview — 46

	3.5.3	Levenshtein analysis of Mazatec dialects	47
	3.5.4	Levenshtein distances for a restricted sample: Nouns	52
	3.5.5	Minimum spanning tree	54
	3.5.6	Dendrogram	55
	3.5.7	Multidimensional scaling	56
3.6	Basque Regional Variation as a Case Study		57
	3.6.1	Database and method	57
	3.6.2	Morpho-(phono-)logical rules	59
	3.6.3	Dialect network from the Levenshtein distances	61
	3.6.4	Concluding remarks on the Basque analysis	65
3.7	Case Study: The Tseltal Diasystem		66
	3.7.1	Subdivision of the Tseltal Dialects	67
	3.7.2	Analysis of the Tseltal database: Morphology	68
	3.7.3	Analysis of the Tseltal database: Phonology	73
	3.7.4	Analysis of the Tseltal database: Lexical variation	74
	3.7.5	Analysis of the Tseltal database: Comparison	76

4 Historical Glottometry — 79

4.1	Background		79
4.2	Wave Theory		80
4.3	Case Study of Numic		82
	4.3.1	The Numic spread hypothesis	82
	4.3.2	The Numic database	84
	4.3.3	Results from historical glottometry	85
	4.3.4	Results from Levenshtein distances	88
	4.3.5	Results from the nearest neighbor map	90
	4.3.6	Results from the dendrogram	91
	4.3.7	Minimum spanning tree	92
	4.3.8	Multidimensional scaling	92
	4.3.9	A short comparison	93

PART TWO: LANGUAGE DYNAMICS

5 Introduction to Language Dynamics — 97

5.1	Motivations behind Language Dynamics Modeling		97
5.2	Numerical Calculations and Experiments		98
	5.2.1	Constructing a model of language dynamics	100
5.3	A Minimum Dictionary		102

6 Language Evolution Models — 110

6.1	Semiotic Dynamics models		110
	6.1.1	The Nowak model	110
	6.1.2	Naming game models	112

6.2	Genetic-like Models	116
	6.2.1 Statistical properties of languages	117
	6.2.2 The Schultze and Stauffer Bit-String model	117
	6.2.3 The Kosmidis, Halley, and Argyrakis Bit-String model	119
	6.2.4 The Viviane De Oliveira model	121
	6.2.5 Modeling the Mazatec expansion	124

7 Language Competition Models — 128

7.1	Introduction to Language Competition	128
7.2	Two-State Models: The Abrams and Strogatz Model	129
7.3	Models with Bilinguals	133
	7.3.1 The Minett and Wang model	134
	7.3.2 Connecting language and opinion dynamics: The AB model	137
	7.3.3 The role of bilinguals	138
	7.3.4 The naming game as a competition model	143
	7.3.5 Similarity between languages: The Mira and Paredes model	145
	7.3.6 Use of language	147
7.4	Models with Population Dynamics	149
	7.4.1 The model of Pinasco and Romanelli	150
	7.4.2 Shared resources: The model of Kandler and Steele	151
7.5	Taking into Account Geography	151
	7.5.1 Inhomogeneous dispersal: Influence of linguistic barriers	153
	7.5.2 Inhomogeneous dispersal: Influence of physical geographical barriers	157
7.6	Wave Front in Language Dynamics	160
	7.6.1 The ecology of languages: The Baggs and Freedman model	161
7.7	Examples of Language Use	164
	7.7.1 Language use in the Basque country	164
	7.7.2 Language use in Catalunya	173

Conclusion — 177

Bibliography — 179
Index — 197

Figures

2.1 Undirected graph with $N = 5, L = 4$, connectivities $k_1 = 2, k_2 = 2, k_3 = 3, k_4 = 1, k_5 = 0$ average degree $\langle k \rangle = 8/5$, clustering coefficients $C_1 = C_2 = C_3 = 1, C_4 = C_5 = 0$.   24

2.2 Erdős–Rényi random graph with $N = 6$, probability $p_0 = 0.7$, expected number of edges $\langle L \rangle = p_0 N(N - 1)/2 = 10.5$.   25

2.3 Small-world network with $N = 24$ (Watts and Strogatz's procedure with rewiring probability $p_1 = 0.4$ from a regular ring with $k = 4$, duplicate edges not allowed).   26

2.4 Scale-free network with $N = 20$ (preferential attachment algorithm with $m = 7$). Some hubs are already clearly visible, despite the small size of the graph.   27

2.5 Phylogram of Mayan languages from the etymological dictionary by Kaufman and Justeson (2003) converted into a database (Léonard, 2014, p. 38). AKA: Akatek, AWA: Awakatek, CHL: Ch′ol, CHJ: Chuj, CHT: Ca a hontal (from Tabasco), CHR: Ch'orti', ITZ: Itzaj, IXL: Ixil, KAQ: Kaqchikel, KCH: K'iche', LAK: Lakandon, MAM: Mam, MCH: Mocho, MOP: Mopan, PCH: Poqomchi', POP: Popti', PQM: Poqomam, QAN: Q'anjob'al, QEQ: q'eqchi', SAK: Sakapultek, SIP: Sipakapek, TEK: Teko, TOJ: Tojolab'al, TZE: Tzeltal, TUZ: Tuzantek, TZO: Tzotzil, TZU: Tz'utujiil, USP: Uspantek, YUK: Yukatek, WAS: wastek (Huastec, or Tének).   36

2.6 Graph of the reflexes of the etymon *jooloom ('head') in Proto-Mayan. From Leonard et al. (2010) (sfs.snv.jussieu.fr).   38

2.7 Cladistic phylogram for Mongolian languages (by Pierre Darlu). Baa: Baarin, Bon: Bonan, Bur: Buriad, Cha: Chahar, Dag: Dagur, Hlh: Khalkha, KJ: Kangjia, Kmn: Kamnigan, Mog: Moghol, Mgr: Monguor, San: Santa, ShY: (Butha) Shira Yugur.   39

3.1 Mazatec dialects considered in our study. The 12 Mazatec dialect locations coincide with those studied by Kirk (1966). (Map data © 2019 Google, INEGI.)   45

3.2 Vitality scale of Mazatec as percentages of Mazatec speakers (see legend). Spots represent urban centers. (CELE [Vittorio dell'Aquila, 2014]. Official census data [2002].) ... 48

3.3 Mazatec dialect networks with normalized Levenshtein distances $L < T = 0.20$ (Map data © 2019 Google, INEGI.) ... 49

3.4 As in Fig. 3.3, but with $L < T = 0.22$ (Map data © 2019 Google, INEGI.) ... 50

3.5 As in Fig. 3.3, but with $L < T = 0.24$. (Map data © 2019 Google, INEGI.) ... 50

3.6 As in Fig. 3.3, but with $L < T = 0.27$. (Map data © 2019 Google, INEGI.) ... 51

3.7 As in Fig. 3.3, but with $L < T = 0.29$ (Map data © 2019 Google, INEGI.) ... 51

3.8 Nearest neighbor network based on the LD distances of 117 cognates, based on the data of Kirk (1966). (Map data © 2019 Google, INEGI.) ... 52

3.9 Dialect networks from average Levenshtein distances between the nouns in Kirk's database, for different thresholds T. (Map data © 2019 Google, INEGI.) ... 53

3.10 Minimum spanning tree based on the Levenshtein distance applied to nouns in Kirk's data. Adapted from Leonard et al. (2017). ... 54

3.11 Levenshtein distance applied to nouns in Kirk's data: Dendrogram. Adapted from Léonard et al. (2017). ... 55

3.12 Two-dimensional projection from multidimensional scaling analysis (in linguistic space) of nouns in Kirk's data. Adapted from Léonard et al. (2017). ... 58

3.13 Map of the 94 localities surveyed by the EAS project (Unamuno et al. [2012]. Marble Virtual Globe: A World Atlas. KDE Platform [http://edu.kde.org/marble].) ... 58

3.14 Non-weighted dialect network from average Levenshtein distances $L_{ij} < 1.8$ (top panel) and $L_{ij} < 2.0$ (bottom panel). (Marble Virtual Globe: A World Atlas. KDE Platform [http://edu.kde.org/ marble].) ... 62

3.15 Weighted network of Basque dialects from pondered average Levenshtein distances $L_{ij}^{(w)} < 1.6$ (top panel) and $L_{ij}^{(w)} < 2$ (bottom panel). (Marble Virtual Globe: A World Atlas. KDE Platform [http://edu.kde.org/marble].) ... 64

3.16 Tseltal localities surveyed in the ALTO project. From http://alisto.org. ... 65

3.17 Original map of the dialect areas, the Greater Tseltalan. From Campbell (1987). ... 66

3.18 Maps obtained from the morphological similarity matrix for three different values of the threshold w [normalized in the interval (0,1)]. ... 70

3.19 Networks of Tseltal varieties according to phonological clustering of pondered similarity index. The thresholds are normalized in the interval (0,1). ... 74

3.20 Networks of Tseltal varieties according to lexical clustering of pondered similarity index [here the threshold is normalized in the interval (0,1)]. ... 76

4.1 Example of intersection of the subgroups A, B, and C. From Rannap (2017). ... 80

4.2 Two examples of intersections of subgroups. Left panel: A, B, C, with subgroupinesses $s_{AB} = 8$, $s_{AC} = 0.88$, $s_{BC} = 0.22$. Right panel: A, B, C, and D, with $s_{ABC} = 8$, $s_{AC} = 0.88$, $s_{AD} = 0.68$, $s_{DC} = 0.28$ (right). From Rannap (2017). 81

4.3 Numic area (showing the Great Basin). Co: Comanche, Pn: Panamint, Sh: Shoshoni, Ka: Kawaiisu, SP: Southern Paiute, Ut: Ute, Mo: Mono, NP: Northern Paiute. (From Rannap [2017]. Map data © 2019 Google, INEGI.) 82

4.4 Tree of Numic languages illustrating the composition of the Western, Central, and Southern groups. From Rannap (2017). 83

4.5 Glottometric diagram of the Numic languages, realized using the 24 strongest links. From Rannap (2017). 87

4.6 Graph of the Levenshtein matrix of the Numic languages database with all the links. (Gephi [gephi.org]. From Rannap [2017].) 88

4.7 Graphs of the Numic database for different thresholds T. (Gephi [gephi.org]. From Rannap [2017].) 89

4.8 Nearest neighbor network of Numic languages: each language is linked only to its closest neighbors. (From Rannap [2017]. Map data © 2019 Google, INEGI.) 91

4.9 Dendrogram of Numic languages. From Rannap (2017). 91

4.10 Minimum spanning tree of the Numic languages. From Rannap (2017). 92

4.11 Result of the multidimensional scaling on a plane. From Rannap (2017). 93

6.1 Examples of communication failure (top panels) and success (bottom panels) in the naming game, in which the speaker conveys word B to the hearer. See text for details. 113

6.2 Average N_w, N_d, and S (1200 simulations, 10^5 time steps, $N = 1000$ agents. From Marchetti et al. [2020]). 114

6.3 Success rate S averaged over 1200 runs versus rescaled time $t/t_{s=0.5}$ ($N = 50$, 100, 500, 1000, 1500, 2000. From Marchetti et al. [2020]). 115

6.4 Star-shaped geographical distribution of languages. Reprinted from De Oliveira et al. (2006b) with permission from Elsevier. 122

6.5 Area–diversity relation. Reprinted from De Oliveira et al. (2006b) with permission from Elsevier. 123

6.6 Enlarged terrain map of the Mazatec area showing some valleys connecting the artificial lake in the Lowlands to the Highlands. (Map data © 2019 Google, INEGI.) 124

6.7 Two-dimensional lattice used in the simulation of the Mazatec expansion. 125

6.8 Time evolution of a simulation of the Mazatec expansion. Each different gray tonality corresponds to a different language. 127

7.1 Scheme of a two-state model of language competition between languages X and Y. The quantities P represent transition rates. 130

7.2 Phase plane of the AS model. The representative point $(x(t), y(t))$ moves along the line $x + y = 1$ ($x, y \in (0, 1)$). For $a > 1 (a < 1)(x^*, y^*)$ is an unstable (stable) equilibrium point. 133

7.3 Scheme of a three-state language competition model, in which direct $X \leftrightarrow Y$ transitions are not allowed. The quantities P represent transition rates. 134

7.4 MW model ($k_{ZX} = 0.4, k_{ZY} = 0.5, k_{YZ} = 0.8, k_{XZ} = 1$). Upper panel: Phase plane with two examples of trajectories. Lower panel: Population sizes versus time for the lower trajectory. Reprinted from Heinsalu et al. (2014) with permission of World Scientific. 136

7.5 Generalized MW model with rates given by Eqs. (7.14): Phase portraits and sample trajectories for four different final scenarios. Reprinted from Heinsalu et al. (2014) with permission from World Scientific. 140

7.6 Population sizes $x(t), y(t), z(t)$ for the lowest trajectory in Panel (a) of Figure 7.5. The asymptotic state is a stable bilingual community. Reprinted from Heinsalu et al. (2014) with permission from World Scientific. 141

7.7 Phase portrait and trajectories for the model of Eqs. (7.15). The k-parameters and the critical values $\alpha_{cr} = 0.4$ and $\beta_{cr} = 0.625$ are fixed. Reprinted from Heinsalu et al. (2014) with permission from World Scientific. 142

7.8 Circles signal appearance (light circles) or no appearance (dark circles) of creole languages in the x-y plane; $x = n_M = N_M/(N_B + N_M)$, the fraction of Mulattos; $y = n_B + n_M = (N_M + N_B)/N(N_M + N_B)/(N_M + N_B + N_E)$, the combined fraction of Mulattos and Bozal (M: Mulattos; B: Bozal; E: Europeans. From Tria et al. [2015]). 145

7.9 Probability densities F_A (lower panels) and F_B (upper panels) of A and B speakers, respectively, at $t = 0$ (left panels), $t = 200$ (central panels), $t = 10^4$ (right panels), for $a = 1.3$, $s_A = 0.52 = 1 - s_B$, and $D = 1$. Reprinted from Patriarca and Leppänen (2004) with permission from Elsevier. 154

7.10 Population densities F_A and F_B at equilibrium for the model of Kandler and Steele (2008), Eqs. (7.27), with space-dependent transition rate $c(x)$. 156

7.11 Top panel: Map of three islands. Lower panel: Potential $U(x, y)$ used in the Brownian motion model (7.29). Reprinted from Patriarca and Heinsalu (2009) with permission from Elsevier. 158

7.12 Population fractions of speakers A and B versus time obtained from the solutions of Eqs. (7.29) in the whole system (top panel) and in each island (bottom panel). Reprinted from Patriarca and Heinsalu (2009) with permission from Elsevier. 159

7.13 Use of Basque with respect to Spanish among the population of the Basque Country, with age ≥ 16 years, by the scope of use. 168

Tables

2.1 Greenbergian morphological types in 13 languages, compiled from the data by Krupa (1966), Greenberg (1960), and Cowgill (2000), on the basis of 100-word samples. ... 33

3.1 Mazatec locations and abbreviations. Parts of names sometimes omitted are in square brackets [·]; alternative names are in parentheses (·). ... 46

3.2 Regions of the Mazatec dialects. ... 47

3.3 Levenshtein distances between pairs of Mazatec dialects for the 12 locations studied by Kirk (1966). (Data processing: CELE, V. dell'Aquila [2014].) ... 49

3.4 Levenshtein distance matrix based on nouns from Kirk's database: 311 nouns overall (Kirk, 1966). ... 53

3.5 Main operations according to the SFVL method. ... 59

3.6 Examples of structural change correlates in the SFVL model: naiz and zarete. ... 60

3.7 Structural change correlates according to the SFVL model. ... 61

3.8 An SFVL-based grid for pondering the Levenshtein distance. Notice the basic hierarchy S≪F≪V≪L. ... 63

3.9 Tseltal dialects of the analysis and respective abbreviations. ... 67

3.10 Morphosyntactic variables. ... 69

3.11 Indexation of the 21 morphosyntactic variables. ... 70

3.12 Similarity index from morphosyntactic data. ... 71

3.13 Phonological variables. ... 73

3.14 Pondered matrix for phonological similarity [value normalized as percentages in (0,100)]. ... 74

3.15 Pondered lexical variables. ... 75

3.16 Similarity indexes [normalized as percentages in (0,100)] from the pondered lexical variables. ... 75

4.1	Number of cognates in the languages of the Numic database, 292 cognates in total.	85
4.2	Examples of irregular innovations.	85
4.3	Examples of regular and irregular innovations.	86
4.4	Ten strongest subgroups of Numic dialects.	86
4.5	Matrix of weighted Levenshtein distances for eight Numic dialects.	88
5.1	Caetano Berruto's typology of linguistic repertoires (Berruto, 2004).	106
7.1	Percentages of users of Basque (with respect to Spanish) with age ≥ 16 years for different scopes and years. Source: Basque Government. Department of Education, Political Linguistics and Culture. Sociolinguistic Survey.	165
7.2	Micro-models in the T situation (1991).	166
7.3	Micro-models in T situation (2011).	167
7.4	Use of Basque among the population with age ≥ 16 years by scope of use, according to province, year 2011.	169
7.5	Micro-models by scope of language use, Alava, 2011.	170
7.6	Micro-models by scope of language use, Bizkaia, 2011.	170
7.7	Language in use in the family (Basque speakers only) according to territory and gender in Euskal Herria (Basque country), year 2011. Values in % (absolute values are explicitly indicated). CAB: Basque Autonomous Community; FBC: French Basque Country; tot.: Total; F: Females; M: Males.	171
7.8	Language used with elders in the CAB and the French Basque Country, 2011.	172
7.9	Language used with children in the CAB and the French Basque Country, 2011.	173
7.10	Language used with children in the CAB and the FBC, 2011.	174
7.11	Population aged 15 years and over according to first, identification, and habitual language in Catalunya, Spain, 2013. Source: http://www.idescat.cat/economia/inec?tc=3&id=da01&lang=en.	175
7.12	Sample of population aged 15 years and over according to first, identification, and habitual language in Catalunya, Spain, 2013.	176

Preface

This book provides a short and selective overview of some quantitative methods of complexity theory and statistical mechanics used in the analysis of linguistic databases and in the study of language dynamics, meaning foremost the quantitative study of language spread and evolution in terms of mathematical models. A number of applications have been included in the book to introduce, illustrate, and explain the theoretical methods considered. Most of the methods discussed are imported from various disciplines, in particular, the hard sciences, where they are standard tools, or have been developed into their current state from new fields, as in the paradigmatic example of the theory of *complex networks* developed within the field of *complex systems* theory.

The general aim of the book is to show the possible advantages in predicting and understanding language processes obtained by integrating the knowledge provided by classical linguistic methods, on one hand, with the information extracted by applying mathematical tools of complexity theory, on the other hand. The general message that we intend to communicate is not new: the interdisciplinary interactions described here have already begun at different times and in many ways. These interactions are due in part to the truly interdisciplinary nature of linguistics and, in particular, to its logical framework, representing a fertile terrain for the introduction of new quantitative methods. We are still in the first stages of a real integration between linguistics and complex systems and there is still a long way ahead, full of challenges and problems to be studied, before we reach our ideal.

The exposition of material from the technical point of view is gradual, starting from basic concepts and models and moving toward more complex features of mathematical modeling and data analysis methods. The structure of the book is designed around some general topics, such as the nature of language groups and some linguistic processes, such as language competition and language evolution. We have tried to keep theory linked with applications to real linguistic problems and have avoided technical details, in order to maintain the book at a general or introductory level, suitable both to researchers and advanced students interested in the interdisciplinary field of language dynamics.

This book is an exploratory study, in which we present a heterogeneous set of methods and examples; the assorted samples should give at least an idea of the possibilities and help us take closer steps toward the long-term goal of the integration of classical concepts and methods of linguistics with more recent approaches of complexity theory, a goal difficult but, we believe, not unreachable.

For brevity, we had to make a selection of topics and examples, which unfortunately prevented discussion or only allowed minimal discussion on many interesting studies and models. Furthermore, the topics discussed reflect our experience and therefore may be biased towards our research. Despite this limitation, the concepts and examples presented in the book should provide a general idea of this developing field.

We would like to thank the many people who have contributed to the present monograph, making it possible. In particular, we would like to thank the following:

- Bien Dobui, for carefully reading and critically checking the whole manuscript both at the scientific and stylistic levels.
- Jürgen Rannap and Flore Picard, for a fruitful collaboration in the study of Numic languages, leading to the results reported in Jürgen Rannap's bachelor thesis, 'Mathematical Analysis of Numic Languages', University of Tartu (2007) (https://dspace.ut.ee/handle/10062/57952) and for the permission to reuse material from the thesis in this monograph.
- Gionni Marchetti, for our collaboration on language dynamics and for allowing us to reuse parts of the results in the form of figures and material from various sections, in particular those concerning complex networks and the naming game.
- Gilles Polian (Ciesas Sureste, México), for supervising the ALTO/Alisto project on Tseltal dialects; providing the dialectometric results with the help of Alain Polian; and generously providing maps and lists of structural variables for Tseltal diasystemic analysis.
- Kiran Sharma and Anirban Chakraborti, for the study carried out in collaboration on Mazatec languages, which represents a relevant case study discussed in various chapters of this monograph, and for the permission to reuse parts of the results of that work.
- Pierre Darlu (Inserm, Paris), for our collaboration in the analysis of the Basque language, and the cladistic analysis of Mayan and Mongolian languages; and for the permission to reuse the results.
- Lyle Campbell and the editors of *Anthropological Linguistics* for kindly allowing us to reprint the original map in Figure 3.17 of the Tseltal area drawn by Campbell (1987).
- MP also thanks X. Castelló, J. Uriarte, V. Eguíluz, and M. San Miguel, for the invaluable experience of our collaboration while writing the article 'Three-state models of language dynamics' (Patriarca et al., 2012).
- Finally, we thank all the other people who kindly agreed to let us reproduce figures from articles they authored and the publishing companies of the respective articles.

The work for this monograph has been carried out by the authors within the framework of some research projects, whose financial support is gratefully acknowledged:

- Jean Léo Léonard thanks the French National Research Agency for letting him be part of the program *Investissements d'Avenir* ANR-10-LABX-0083 (Labex EFL), in particular PPC11–Complex Systems & Diffusion of Phonological Traits (strand 1) and Cross-Mediated Elicitation, EM2 (strand 7).

- Els Heinsalu and Marco Patriarca acknowledge the support of the Estonian Ministry of Education and Research, Institutional Research Funding IUT (IUT39-1) and the Estonian Research Council, Grant PUT (PUT1356), and ERDF (European Development Research Fund) CoE (Center of Excellence) Program through Grant TK133.
- We acknowledge the support of the Hubert Curien Foundation, within the Estonian-French Science and Technology Cooperation Program, Parrot, through the project MOLDIC (Modeling Language and Dialects with Complexity Theory), 2017–18.

Finally, we would like to thank the editors of Cambridge University Press, who carefully followed the development of the work.

Some of the maps in the book were realized with the invaluable help of software such as Google 'MyMaps' and Marble (KDE).

Marco Patriarca[*], Els Heinsalu[*], and Jean Léo Léonard[^]
Tallinn, March 6, 2020

---

[*] NICPB – National Institute of Chemical Physics and Biophysics, Tallinn.

[^] Sorbonne Université, STIH (EA 4509) (2014–2019); Université de Montpellier III (Paul-Valéry), Dipralang (EA 739) (2019–).

Chapter 1

Introduction

1.1 Introduction to Complexity

Complexity theory is a major interdisciplinary paradigm which unifies natural and social sciences through a combination of quantitative and qualitative methods applied at various phases of the research, from observations and data analysis to modeling, simulation, and interpretation of complex phenomena (Anderson, 1972; Ross and Arkin, 2009). In this framework, cognitive and nonlinear stochastic models and effective representation methods such as complex networks and fractal geometry, represent a part of the standard toolbox. In fact, in complexity theory, phenomena emerge dynamically from hierarchical, multi-modular systems, produced by bundles of possibly stochastic interactions and causalities rather than from correlative determinism.

As far as language and linguistic analysis are concerned, complexity itself can be understood from at least two different perspectives. On the one hand, there is 'constitutional complexity', or 'bit complexity', that is, complexity due to inventories of functional units or structural features, such as phonemes, morphemes, and lexical stems. On the other hand, there is '(socio-)interactional complexity', or, in other words, 'communal complexity', involving intricate modules of units and features, or networks of interactive individuals and aggregates. These different aspects all find their natural description in the *multiplex* paradigm, that is, through a model system composed of a set of interacting, overlapping networks. The unification of the two aforementioned dimensions of complexity represents a major challenge and is a focus point of this book.

For more than a decade, a growing interdisciplinary community has applied the tools of complex systems theory and statistical mechanics to the study of problems that traditionally belong to the field of linguistics. Nowadays, language dynamics represents a relevant branch of complexity theory. The modeling of language dynamics has mainly addressed three fundamental dimensions of language complexity.

(i) **Language spread and competition** (the dynamics of language use in multilingual communities).

(ii) **Language evolution** (how the structure of language evolves).

(iii) **Language cognition** (the way the human brain processes linguistic knowledge).

While these three dimensions of language complexity closely interact with each other and should all be taken into account for an exhaustive description of language complexity, it is useful, for clarity, to consider them as separate aspects. In the present book, we mainly address the first two dimensions, discussing language spread and competition models and considering language evolution models. However, we will also use socio-cognitive models of linguistic and cultural change.

A fundamental characteristic of complex systems theory is its high interdisciplinarity: the methods and the models used share many relevant features with other disciplines. This is simply due to the fact that, by definition, the paradigms of complex systems theory apply to all complex systems, independently of their specific nature. In the following, we present a short list of common points between the complex systems approach to language dynamics and other disciplines.

First, language dynamics is clearly connected to problems related to social interactions. Such a direct connection with social sciences and social dynamics follows from the mere fact that linguistic features represent cultural traits of a specific nature, namely, semiotic traits, distributed on two levels of organization: as morphemes, words, and phrases on the one hand; distinctive features and the so-called auto-segmental traits on the other hand. The propagation and evolution of these complex, double-sided items (words versus traits) can in principle be modeled through dynamics analogous to those of cultural spread and opinion dynamics. However, one should also consider that the conditions of their spreading in time and space may turn out to be more intricate, as they resort more to indexical traits than mere artifacts, opinions, or ideas. In other words, their liability is higher than for artifacts. Moreover, the complexity of diffusion and equilibrium between competing languages or dialects differs strongly according to the patterns and properties shared among languages: the more akin languages are, the more they mingle through contact; whereas, genetically and typologically distant languages undergo stiffer processes.

Many models considered in the book are models of competition between cultural traits and are usually referred to as 'language competition models'. Even when language change and evolution processes are ignored, language competition models can offer the invaluable opportunity to understand the mechanisms regulating the size evolution of linguistic communities, representing an important part of the information that needs to be taken into account, for example, in the design of appropriate linguistic policies. Language competition is a central concept in this book and reveals another relevant analogy, namely between language dynamics and the dynamics of ecological systems. This analogy can be particularly useful for representing in a simple way the more complex landscape of interactions between the individuals of two different-speaking communities. Interestingly, such an analogy seems to suggest that single languages (or linguistic traits) are the actual entities competing with each other for resources, while speakers are the 'mere niches' where selection and evolution eventually take place. The promotion of languages to the role of main characters recalls the definition of 'meme' introduced by Dawkins (1976), in this case, with memes representing languages or linguistic traits competing with each other for resources in the minds of speakers, analogously to the way genes compete for resources at the genetic level. This type of framework is certainly useful when considering minimal models of language competition, where the relevant variables are the sizes of the various interacting communities, as in a Lotka–Volterra modeling framework. However, even in the perspective where languages interact as competing species, one should bear in mind that a competition process between two languages differs in many respects from competition between species.

(i) First, language is a **cognitive system** or, as Noam Chomsky put it (Chomsky, 1986), a cognitive device, based on universal constraints (principles of universal grammar, for example, markedness versus default, topicality and comment constituent structure of predication, lexicon versus functional heads, etc.) and accommodated to local parameters (for example, accusative versus ergative alignment, head marking versus dependent marking, etc.).

(ii) Second, **social time can evolve much more rapidly than biological time**—and language, as a matter of fact, is embedded in social periods and social time, interconnected with a complex set of cultural and political constraints; see Dixon (1997) for a punctuated model of language evolution and shift.

(iii) Third, language is ontologically a **rather freely articulated system** of rules and constraints, with a very high sensitivity to setting (that is, status of the speakers involved in social intercourse and to the context of communication), plasticity being the rule, rather than the exception, in colloquial interaction.

The first of the factors listed here accounts for the potential complexity of repertoires. The second accounts for complex patterns of emergence, while the third factor for powerful patterns of miscegenation and self-organization. The emergence of creoles provides a good example of the influence of these factors in the evolution of new means of expression and linguistic patterns leading to totally new languages. The first factor accounts for universal trends such as low markedness and primarization of complex structures, and local parameters, such as right-branching determination (as in Haitian Creole) or left-branching determination (as in Seselwa or other Mauritian Creoles). The second factor powerfully accelerated the emergence of creoles in just a few generations, in spite of the appalling social violence enforced or enslavement of human aggregates. The third factor expands the complexity and the plasticity of the repertoire, from basilect to acrolect (see further for a definition of these terms).

Before proceeding further, it would be good to clarify that the complex systems approach developed in this volume differs conspicuously from the so-called computational linguistics or automated text analysis, whose purpose is to process linguistic data or corpora in order to implement grammatical models through computing automata, or to widen the empirical range of analysis with powerful tools for processing and parsing data as done by Clark et al. (2010) and Grishman (1986). Computational linguistics and automated text analysis are concerned with algorithmic complexity, as is complexity theory, but the main goal of those approaches is data extraction and a more accurate and comprehensive description and modeling of the fabric of the lexicon, of grammar or discourse. For example, typical research in computational linguistics may attempt to account for the complexity of verb inflection in Bulgarian or Macedonian, or to contrive more accurate and efficient multilingual automatic translators or devices for speech synthesis, etc.

1.2 Aim of the Book

Many of the canonical models that will be discussed in the present book have been developed over more than a decade by physicists, who have applied the tools of statistical mechanics and complex systems to study the problems that traditionally belong to the field of linguistics. The interest among physicists and the complex systems community in modeling language competition was originally raised by the work of Abrams and Strogatz (2003) (see also Wichmann [2008a,b] for a review). However, research on the dynamical modeling of the interaction between linguistic communities had already begun more than a decade earlier with the papers written by Baggs and Freedman (1990, 1993). These papers address the problem from two different sides that are still central in the current research in language dynamics:

(i) A mathematical formulation based on a close analogy with the ecology of competing biological species: thus, language dynamics (or at least the study of language competition) can be considered a natural branching of the mathematical modeling of ecological processes.

(ii) The concern for the problem of the disappearance of language and cultural diversity, similar to its ecological counterpart of the disappearance of species diversity.

In our opinion, it is now the right moment to look back on what has been done so far by the growing interdisciplinary community of researchers working in language dynamics, and to draw some conclusions on the picture obtained by the various studies. This introspection will also allow us to ponder over the most relevant and reasonable research directions to be taken in the near future by applying complexity theory to language dynamics. These questions have not been considered systematically so far. The scientific goal of the present book is to fill this gap, which affects different disciplines, and to provide a reference point for the highly interdisciplinary studies concerning the mathematical modeling of language competition, evolution, and spread. On the social side, as long as the defense of language diversity is a concern, the models studied can be used for planning optimal strategies to defend endangered languages or at least limit what in most cases seems to be an unavoidable imminent cultural catastrophe.

The treatment of mathematical modeling of (socio)linguistic complexity is developed in the present book around the following two lines:

- First, the exploration of a theoretical or virtual range of facts and phenomena that can be applied to language in space and time, from a logical and hypothetico-deductive standpoint.
- Second, the verification of the match between the models and their corresponding results obtained through simulations, on the one hand, and with the corresponding (socio)linguistic data, on the other hand.

The aforementioned two points are closely related to each other. The first point resorts especially to analytical calculations and numerical simulations in complex system modeling as main investigation tools; the second one aims at challenging and validating theoretical models with empirical facts. Various interesting case studies, such as that of the Mazatec dialects or the Basque language, will be employed throughout the text to present applications to real problems that illustrate concepts and methods: language ecology, linguistic evolution, structural complexity, and the various computing tools employed.

Particular attention will be devoted to the (in)adequacy of abstract models of language diffusion in space and time, or language competition in contexts of diglossia (subordinate bilingualism or bidialectalism) or language planning, highlighting the relevance of (socio-)linguistic facts from other fields not strictly (socio-)linguistic.

The goal of these types of studies been clear in the complex systems community for a while. The need to make social dynamics, in general, a more 'sound discipline grounded on empirical evidence' has been well explained, for example, by Castellano et al. (2009) and Taagepera (2008). In other words, we are looking, in the field of language study, for the standard procedure common to any scientific research, namely, the development of a set of models and laws which are applicable within well-defined ranges of validity. In fact, considering the possible applications of this method, the approach advocated in this book has a strong predictive potential on the future trends of linguistic communities and the evaluation of the state of endangered languages, or for shaping the choice of effective language policies to protect minority languages.

Furthermore, the models developed can also be reapplied to interpret facts with models and algorithms, that is, to analyze new data sets in light of the phenomenological models validated, in order to test historical hypotheses and the actual mechanisms of cultural spread. In questions of a historical character, predictions based on an initial or an abstract state need to be checked by facts; conjuncture should sustain conjecture.

Our approach also aims at filling a gap between advanced research in language and complex systems and routines already deeply rooted in both fields of expertise. On the one hand, in hard sciences, one often relies too much on abstract models, providing results considered as irrelevant by linguists. On the other hand, linguists observe intricate situations of multilingualism and linguistic diffusion, without being conscious of the weight and consequences of micro-trends and parameters. Mathematical modeling can help to conceptualize and deliberate on empirical facts and situations better than from a mere ethnographic standpoint, while the pragmatic interpretation of facts and figures through sociolinguistics may considerably foster mathematical modeling in its endeavor to predict and represent reality from as broad a scope as possible. Even if we do not fill this gap, we hope that, by putting the different approaches side by side and using them to study some real problems, we are making further steps toward a constructive integration of the different methods and viewpoints.

1.3 The Readership

While the topics and goals of this book can be easily situated within the wide and new research field of complex systems theory, it may seem that this is another instance of physics trying to invade other disciplines. Physics has been undergoing deep, symbiotic interdisciplinary developments that address many problems of biology (biophysics), economics (econophysics), sociology (socio-physics), ecology, and now linguistics as well, resulting in interdisciplinary approaches that are sometimes criticized, as in the case of econophysics.

However, the case of linguistics is different for many reasons. In fact, besides the physicists interested in linguistic problems, there is also a growing part of the linguistic community that has recently become aware of the possibilities offered by complex systems theory. Therefore, the target audience of this book includes those physicists willing to discover the ongoing problems in linguistics that one could address using the theory of complex systems and linguists willing to learn the basics of complex systems theory in order to communicate and collaborate with the former. In order to favor productive interaction between these two communities in the future, it is important to first learn the other's language, that is, the terminology, as well as the corresponding methods and specific ways to approach a problem. For these reasons, the book is written in a way that also makes it accessible to non-specialists, for example, graduate and undergraduate students of both fields in general. Thus, this book may be used in academic courses on complex systems, social dynamics, social sciences, ecological linguistics, quantitative linguistics, mathematical modeling, etc.

1.4 Structure of the Book

The book is organized into two parts. The first part may look more 'linguistic', being concerned about the analysis of linguistic databases, while the second part may look more 'mathematical', being more focused on the modeling of language dynamics. In fact, both the parts have a deeply linguistic side and require the formalization of some quantitative method.

Part One concerns the problem of interpreting linguistic databases to reveal the presence of possibly related languages and determine their main features. This is a relevant side of the approach presented in this book since linguistic databases provide the empirical comparison terms for validating quantitative methods and mathematical models. Part One also contains basic notions about language change and how to quantify the degree of relation between languages, about complex networks, and about techniques based on Levenshtein distance. Their use is illustrated with the help of some working examples, including the Mazatec, Basque, Tseltal, and the Numic languages. Some of these examples come from our own research, started in 2008 as an informal, interdisciplinary study by a group of mathematicians, physicists, and linguists who met at a symposium in Tartu.

Part Two deals with the modeling of linguistic systems, that is, with the actual mathematical models developed so far. Various types of models are discussed, ranging from microscopic individual-based models to macroscopic continuous models of language evolution, competition, and diffusion. The overview is neither exhaustive nor systematic—many interesting models are not discussed—but tries to provide an outline of the field through a selection of some minimal models.

PART ONE
REVEALING LANGUAGE GROUPS

Chapter 2
Language and Languages

While investigating a group of languages, one is interested in formulating an insightful picture of their internal structure as well as of the relations between the different languages in the group, reflecting the bit complexity and the social complexity aspects discussed briefly in the previous chapter. In order to proceed to a comparison and analysis of the languages, one needs to know how to define the differences between languages and group different languages together in a way that best suits the observed differences. From a more general viewpoint, the picture thus obtained cannot be disentangled from the processes underlying the origin and evolution of languages, possibly from a common ancestor. These and other related topics will be discussed in this part of the book.

Unless one is specifically interested in the comparison between different scripts, language comparison methods are not applied to written texts but to the actual spoken languages, by analyzing their phonetic transcriptions into the IPA, the International Phonetic Alphabet[1]

The transcription of a language into IPA is difficult work, requiring the field linguists to make a faithful transcription of the spoken language even up to its smallest phonetic details. In the case of unwritten languages, that is, languages that do not have a standard written form used by the native speakers, their IPA transcription represents the only existing written source.

One of the relevant goals of an analysis of a linguistic database is to provide a reliable picture of related languages. Some prototypical linguistic scenarios used to describe language groups are the following:

- **Linkage**. A linkage of languages is a group of languages deriving from a protolanguage that has gradually branched into a number of (still) interacting languages. However, despite the continuous interactions, such languages keep differentiating more and more into separate languages. The term 'linkage' was introduced by Ross in his study of Western Oceanic languages (see Ross [1988]). A typical example of linkage is a group of languages born within a *dialect continuum*. The linkage formation process resembles

---

[1] The IPA, International Phonetic Alphabet, is a notational standard for the phonetic representation of all languages, maintained by the International Phonetic Association. It is summarized in the famous *IPA chart*. (For an interactive version of the IPA chart, see www.internationalphoneticassociation.org)

parapatric speciation in ecology, where species differentiate while remaining in contact with each other.
- **Family**. In a so-called *family* of languages, the languages derive from the protolanguage and at certain points, they separate into groups, which then remain isolated from each other, rather than forming an interacting network of speech communtites. Language family formation has certain similarities with *allopatric speciation*, where different species diverge genetically from the common ancestor, after being geographically separated from each other.
- **Sprachbund** (or *linguistic area*). A group of languages, not necessarily deriving from a common ancestor, may eventually form a sprachbund after acquiring some common linguistic features from each other by influencing each other during their history. The languages in a sprachbund thus have some features clearly common with each other and additional features that set them definitely apart. The formation of a sprachbund finds a genetic analogy in the more general situation in which both horizontal and vertical gene transfer can take place as, for example, in populations of bacteria.

The reconstruction of the history of a linguistic landscape can be extremely complex, due to the fact that the borders between the aforementioned categories are not sharp; in some real situations, it is difficult to categorize the system according to just one definition. An analysis implicitly makes some assumptions, which may polarize the final conclusions, for example, the existence of a genetic tree. Therefore, for the analysis of a group of languages to reveal the actual nature of their mutual relationships, the choice of a suitable estimator of the (dis)similarity between the languages has a crucial role.

In this chapter, we will consider some general methods and examples for the general problem of the evaluation of the differences between languages and the structure of their connections. In statistical learning, a problem of this type, in which one tries to infer the underlying structure and patterns of a system from the observed data, is referred to as a problem of unsupervised learning (Hastie et al., 2009), while the complementary problem, in which one tries, for example, to infer the behavior of a system from a given parameter set (dynamical parameters and initial conditions) is a problem of *supervised learning* and is the typical problem encountered in the numerical simulation of a system.

When dealing with languages, methods based on string metrics are the most straightforward and intuitive approaches to language comparison through the IPA database. Examples of methods of unsupervised learning are principal components analysis and multidimensional scaling, which attempt to identify some equivalent low-dimensional manifolds, or the family of cluster analysis methods. Some of these methods are illustrated in the following sections, together with the presentation of a corresponding case study. Certain methods partition the languages into disjoint subsets, for example, the hard-clustering methods, while others, the fuzzy clustering algorithms, allow a data point to belong to more than one cluster with different probabilities. The latter methods are suitable to interpret data according to a more general theoretical framework, such as wave theory, as in the case of historical glottometry, which is discussed in Chapter 4.

2.1 Language Evolution

Similar problems are studied in other fields, from which useful concepts, methods, and algorithms can be imported into the study of language.

Evolution is a general process characterizing complex systems; it can change a system both at the basic level of its microscopic features and in the macroscopic level of its emergent properties, as in the case of the so-called *complex adaptive systems*: examples are biological systems and language. Due to its universal character, evolution is studied in various fields, from ecology and biology to artificial intelligence, economics and other social sciences. In this section, we will try to connect the evolution paradigm to language and language changes.

An essential ingredient of any evolutionary process is the presence of random mutations that take place in a unit of the system and possibly affect various features of the mutated unit. When transmitted to the next generation or transferred to other units, mutations act as the random seeds of new complex evolutionary processes. However, the complex shape of the evolution process results from the interplay between many elements, that is, between mutations, competition and natural selection, and dispersal; the evolutionary processes observed have an overall pluralistic nature due to their dependencies on additional factors, effects, and nonadaptationist mechanisms, such as genetic drift, ecological and historical constraints, internal dynamics and symmetry constraints of the system, and exaptation (the new use of parts originally adapted to a different function).

As far as language is concerned, 'evolution' can actually refer to very different, possibly related, processes (Hurford, 1994; Nowak, 2000).

- First, language evolution could refer to the evolution of particular languages, for example, the differentiation of Latin into the Romance languages, a process taking place on typical time scales of a few thousand years, during which one can see new languages appearing. This same time scale also characterizes the time span in which it is possible to make a linguistic reconstruction of a protolanguage.[2] This type of evolution concerns various parts of language and the corresponding changes can be grouped into changes of phonetics, phonology syntax, and semantics.
- 'Language evolution' also includes the evolution of the language faculty, that is, the development of human capabilities making it possible to produce more and more complex languages, assumed to have developed from a system similar to that of animal communication (Christiansen and Kirby, 2003; Fitch, 2010; Hurford, 2014). In this sense, language evolution is still an open and debated problem: neo-Darwinians and supporters of non selective viewpoints have opposing arguments and defend different interpretations of the nature of language evolution. The evolution of the language faculty is a process taking place on the time scale of at least hundreds of thousands of years, considering the time scale for the appearance of *Homo Sapiens*, but actually it is not known how developed previous languages may have been. During this period, human

---

[2] A protolanguage is an ancestral language from which it is assumed that many languages were derived (Rowe and Levine, 2015).

language capabilities and language complexity must have evolved together. The outcome of such a process is a complex language that contains a double level of complexity, referred to as the 'duality of patterning', that is, (a) how strings are combined together to form phonemes and (b) how words can form sentences: language makes use of combinatorics on these two levels of complexity, providing language with its wide capabilities.

- Furthermore, a proper biological basis must be present to provide humans with the practical conditions for the use of language. The time scale to develop such a biological premise may be distinct from (and even longer than) the one during which humans started to speak, gradually developing a complex language as an emerging cultural artifact. The evolution of a suitable biological framework has probably proceeded gradually, given the complexity of the speaking and hearing processes: at a physiological level, the vocal tract performs movements that are accurate to within a few millimeters and synchronized to within milliseconds (Miller, 1981); this also implies the presence of a strong exaptation process, given that organs originally devoted to other goals are used for language purposes. Furthermore, a combination of specific physiological and cognitive abilities is needed for language perception and the elaboration of the corresponding information, which depends, in particular, on the way specialized neurons function and the mind as a whole manages linguistic processes (Miller, 1970; Lieberman, 1984). The details concerning the underlying mechanisms and, in particular, the coevolution of biological features and cognitive abilities still remain a puzzling question today.
- In parallel, the pragmatic dimension of the evolution of language, closely connected to the history and ecology of humans, should be considered. The use of language for different goals and activities may have influenced language development and vice versa in a most interesting and constructive evolutionary loop. In other words, the pragmatic and social characters of language may have been the basic elements in shaping language. In connection, one can observe that the evolution of language, in all its complexity, remains a specifically human phenomenon, currently not found in other species, just as it is the case for the evolution of scientific and technological innovations. This may be more than a coincidence and could be due to the fact that language and collaborative habits may have evolved in parallel. Several facts now support this pragmatic view of language origin. For example, in the period between 45,000 and 40,000 years ago, a remarkable advance in human employed technology took place. Furthermore, recent fieldwork experiments, which reconstructed working activities that employed stones as the raw material, have demonstrated that the ability of humans to communicate and exchange information about elaborate concepts with each other is a necessary condition for carrying out certain tasks and therefore for technological development (Cataldo et al., 2018; Faisal et al., 2010; Morgan, 2015).

In the last decades, these and other issues regarding language evolution have been steadily on the rise as topics of study; hence, the reader will easily find a considerable bulk of research, results, and insights on the cognitive, social, systemic and historical aspects of language. In particular, we mention John Wendel's view of the ecology of language (Wendel, 2005), which can be described in terms of (a) diversity, (b) variation, (c) transmission, (d) lineage,

(e) selection and adaptation (or exaptation, as secondary transformations from an initial adaptive asset), (f) niche, (g) rates of change. Wendel illustrates the analogy between language and species through a comparative table, showing how the aforementioned properties match for both species, in nature, and language, in society (see Wendel's paper [2005] for details).

Nevertheless, the definitions of 'language family', 'language death', 'life of languages', etc., and the various comparisons made between language and species are but metaphors, whose value is limited, despite the fact that, as all metaphors, they help us to grasp complex phenomena and issues. For instance, the dialectologist and semantician, Mario Alinei, questions the validity of most of our current views on language evolution from a Darwinian standpoint, noting that rapid changes are a superficial byproduct of intensive social interaction, whereas deep changes evolve slowly in language (Alinei [2006]).

A comprehensive approach to the many issues linked with this field of research can be found in Mufwene (2008), which deals with natural selection in languages, evolutionary trends and the actuation question, multilingualism and language contact, rise and fall of languages, the emergence of creoles and pidgins, impact of colonization and globalization on language vitality, among others. Furthermore, McMahon and McMahon (2013) provide detailed insights into a wide range of issues related to the cognitive dimension of evolutionary linguistics, language in the brain, language and genes, saltation vs. gradual change, reconstruction of protolanguages, and computer simulations of emerging languages (also see Tallerman and Gibson [2012] for a handbook in the same vein). Among the research groups involved in research along these lines, we mention the research activity of the ASLAN[3] consortium, which is also concerned with complexity in language. Finally, we include a most interesting case study of the emergent paradigm of social cognitive models of language change which is based on the interaction between exemplars (tokens, data), form–function reanalysis, form–meaning mapping, and iterated learning (see Hruschka et al. [2009]).

2.2 Language Change

In this section, we limit ourselves to some questions of language change relevant to the scope of this book, with a focus on the following questions.

Do linguistic systems exist as discrete entities, or are they emerging structures, the result of multiple internal constraints and external factors? How do languages cluster phylogenetically, but also expand geographically in space and time, as inalienable—although flexible and permeable—attributes of the populations using them under various circumstances and situations (hence, the choice of stylistic register, and the practice of code switching, etc.)? Indeed, languages are adaptive systems (see the language dynamics model, explained later, applied to Mazatec diffusion and evolution) and the set of sound changes, for instance, though wide, is limited and highly constrained by multiple factors: mechanic or cognitive (see later in this section). As any other biological but also cultural phenomena, language, when expanding

---

[3] The ASLAN (Advanced Studies on Language Complexity) consortium is headed by François Pellegrino (DDL, until 2011), Andrée Tiberghien (ICAR, until 2014), and Matthieu Quignard (ICAR); see http://aslan.universite-lyon.fr/ for details.

in space and time, undergoes a strong double pressure: on the one hand, it reproduces its heritage (lexicon and grammatical units or items), but on the other hand, it innovates from generation to generation and absorbs features and items of neighboring systems with which it enters into contact. Any language is thus both the innovative product of its own genetic pool (its genus within a linguistic family) and the reflex of many contacts with neighboring linguistic systems. No language evolves in complete isolation: even the most remote language of the most isolated tribe comes from somewhere and has been subject to intercourse, at least during its progression to its final isolation point. Moreover, it is well known that languages considered as phylogenetic isolates, clearly show traces of stratified contact over the ages: Basque, in south-western Europe, shows an impressive number of Latin and Romance loanwords in its lexicon; Huave (Ombeayiiüts), the language spoken on the Isthmus of Tehuantepec in Mesoamerica, has unassailable signs of multiple contact with other Mesoamerican linguistic genera or linguistic stocks, such as Western Mayan, Zapotecan (Oto-Manguean), Mixe–Zoque, etc. Languages are as much sponges as viruses, from the standpoint of adaptation to their geographical and social context, and from the standpoint of expansion in space and time. They are also strongly prone to exaptation, though metatypy (what we will call loop, one of the basic mechanisms of sound and grammatical change) is also common. Metatypy is a change in morphosyntactic type and grammatical organization of a language induced by speakers' bilingualism and driven by grammatical *calquing*, the copying of constructional meanings from the modified language and the innovation of new structures using inherited material to express them.

Adaptation copes with what surrounds a communal aggregate, whereas exaptation converts adaptive products into innovative solutions for survival, expansion or, more simply, for ergonomics. Both trends compete in language evolution – a good example of exaptation is what Kartvelian languages, in Southern Caucasus, did with number agreement in verbs (see Tuite [1988]); the same could be said about the emergence of a definite conjugation paradigm in Mordvin, where a reinterpretation of the function of plural markers in rules of stem choices and syncretism plays a central role (Keresztes, 1999, pp. 69–74). The whole bulk of grammaticalization of lexical items into functional markers such as tense, aspect, and mood in creole languages can be viewed as a process of exaptation—to the point that languages can even be considered as more exaptive systems than merely adaptive ones. This points to a characteristic property of linguistic systems, that differentiates them from biological systems: their high potential for structural, but also categorical plasticity, which creates conditions for intensive mingling. Mixed languages, such as Michif-cree, in Canada and the USA, and, to some extent, the Maltese language (a Semitic language incorporating many lexical and grammatical elements from Italo-Romance dialects, especially Sicilian), are good examples of this plasticity, one more aspect of the exaptive trend in languages. This change is driven by grammatical calquing. A concomitant of this reorganization of grammatical constructions is often the reorganization or creation of paradigms of grammatical functors. Usually, the language undergoing metatypy (the modified language) is emblematic of its speakers' identity, whilst the language which provides the metatypic model is an intercommunity language. Speakers of the modified language form a sufficiently tight-knit community to be well aware of their separate identity and of their language as a marker of that identity; however, some bilingual speakers, at least, use the intercommunity language so extensively that they are more at home in it than in the emblematic language of the community.

This points to another issue which has to be tackled when considering linguistic evolution through space and time: to what extent are phylogenetic criteria more or less important than typological criteria, and should evolutive linguistics rely more on the latter than on the former, as the whole WALS (World Atlas of Language Structures) project suggests, institutionally embedded as it is within the Max Planck Institute for Evolutionary Anthropology?[4] In the early nineteenth century, the founders of modern linguistics, Franz Bopp, Rasmus Rask, and August Schleicher, favored the former. However, in the last quarter of the century (Morpurgo Davies, 1998), the so-called Neogrammarian paradigm became predominant, with figures like Karl Brugmann, August Leskien, Hermann Osthoff, Hermann Paul, and Eduard Sievers claiming that sound change should be viewed as inexorable and exceptionless apart from the occasional whims of borrowing (adstratum) and analogy (inference from already existing patterns).[5]

Another matter that needs to be considered is whether convergence areas and feature pools should prevail over genera, linguistic stocks, and phyla (Heggarty, 2017)? Or at least, are the former more heuristic than the latter, in terms of this specific and challenging branch of evolutionism that is linguistic evolution? Should diffusion waves and sociopolitical stratigraphy (the vertical strand, substratum, implying subordination to a hegemonic contact language or register vs. superstratum, implying an intrusive and expanding elite imposing the prestige of its language or register, and adstratum, implying more or less horizontal intercourse between neighboring communal aggregates) be the masters of the (evolutive) game (see the discussion in Léonard [2017] and Ostler [2005] about how hegemonies change the world's languages scenery)? Between inheritance and contact, immanence and contingency, who wins? On all these crucial issues of considerable resonance from an epistemological standpoint, complexity theory has much to say, as this volume suggests.

Nevertheless, however labile and flexible linguistic patterns may be, however diversified dialect networks may appear, languages still make up densely consistent systems, which can easily be modeled and analyzed qualitatively (through taxonomies and combinatory constraints) and quantitatively, as we will see in many occasions in this volume. One can even wittily say that languages are so systemic and consistent that they can afford to be structurally porous and practically labile in current practice and social intercourse, through wide arrays of multilingual or stylistic registers (dialects, sociolects, idiolects, genderlects, variable script habits or more or less ad hoc norms within literacy, etc.). One basic condition ensures this robustness: mutual intelligibility, the best safeguard against chaos. Even speech production constraints cannot violate this basic prerequisite, inducing more compact, allusive and easy forms of speech that can be uttered effortlessly in colloquial intercourse: it is possible to attain this parsimony as the taxonomic and combinatory framework is efficient and sufficiently flexible. Languages constantly strive for communication and to maintain or attain the subtle dialectics of opacity in casual speech (or tokens), associated with the transparency of mental

---

[4] See the website http://wals.info/.
[5] See Léonard (2016) for a more nuanced epistemological survey of the Neogrammarian doctrine and its impact on today's linguistics.

states (or patterns), for optimal encoding and decoding patterns from *parole* (speaking) to *langue* (language) and vice versa, making up the very thread of language use. As we will see, complexity theory can, in many ways, fathom and elucidate or help to visualize the tenets of how such an intricate and paradoxical flow of information can nevertheless stir up constant emerging patterns and at the same time, enforce and strengthen self-organizing lexicons and grammars. Language has often been compared with living organisms. Metaphors such as 'life and death' of languages have become common, nowadays. However, these are indeed far-fetched metaphors. If language is to be compared to an organism, it could be more reliably considered as a semiotic 'being' than any 'living' or organic item from the realm of nature. Language, thus, is basically a semiotic system created and passed on, with astonishing flexibility, albeit robustness, from generation to generation, by human societies. Its fundamental ontology is semiotic, and has nothing to do with biology: even when speakers die, or leave its current form, language may survive, as Akkadian or the Mycenaean Greek language do, or thrive, as Cornish or Manx Gaelic do, through revival strategies. Language diffusion even differs conspicuously from cultural diffusion, in terms of range and scope, as recently suggested by Holman et al. (2015): it is structurally (that is, systemically) more consistent and robust, and can expand farther than cultural traits.

2.3 Mechanisms of Sound Change

The history of a language is marked by the changes that the language undergoes due to many reasons related either to its internal dynamics or its interaction with the environment and other languages. We begin with an overview of the main mechanisms responsible for sound change and the methods that have been used to model them.

Main types of sound change

Most sound changes in the world's languages can be categorized into a few simple formulas, which are illustrated here, with a few examples from natural languages.[6] In the following, 'x' and 'y' represent two distinct generic phonems.

- **Assimilation**. – Progressive assimilation, xy > xx (ex.: Lat. fémina > femna > Old French fæmm > Modern French fam, spelled ⟨femme ⟩).
  – Regressive assimilation, xy > yy (ex.: Lat. dòmina > domna > It. donna 'woman').
- **Dissimilation**. xx > xy (*wiṣ-su locative plural > *viṭsu > vikṣu or Lat. arbor > Spanish arbol).
- **Fortition (strengthening)**. x > X (Latin (loan) fiku > Basque biku), x > xx (Uralic *jikæ> Saame jâkke, 'year'), x > x′x (*leeme > liebmâ, 'broth').
- **Weakening or lenition**. X > x (Lat. rípa > ríβa > Fr. [riv] spelled ⟨rive ⟩, 'shore') [p] as an unvoiced stop, being stronger than a voiced fricative, such as [v].
- **Coalescence**. xy > x^y (Lat. álba > áłb > áub > French [ob] spelled ⟨aube⟩, 'dawn').
- **Deletion**. x > ∅ (Lat. uíta > víð > ví > French [vi] spelled ⟨vie ⟩, 'life').

---

[6] A *natural language* is a language that has evolved naturally in humans through use and repetition, without conscious planning or premeditation (in contrast with constructed and formal languages such as those used to write computer codes or in logic).

- **Insertion**. ∅ > x (Lat. flore > Portuguese (dial.) fulor, 'flower').
- **Metathesis**. xy > yx (Lat. kápra > kábr > Gascon kráb).

At the top of the list, assimilation and dissimilation are strongly active throughout the Mazatec dialect network, and account for many differentiation patterns between vowels and within word templates (see Sections 3.5.3 and 3.5.4).

Fortition and weakening are good examples of how these basic phonological processes may be involved in the computation of edit distance and in the framing of results obtained in our series of graph networks for different languages according to specific typological conditions. These two processes determine to a great extent the inner differentiation of Numic languages (Section 4.3).

The last three processes (deletion, insertion, and metathesis) are typical of changes happening in the Basque dialect network (see Section 3.5), and are easy to process and analyze with edit distance algorithms, such as the Levenshtein algorithm. Deletion matches the VOID variable, whereas insertion matches the FEED. Metathesis typically belongs to the LOOP family of rules, as sounds loop over the word (see Tables 3.5–3.8 in Section 3.6).

The examples given here are but a few instances of how these basic mechanisms work according to typological constraints in languages and dialect networks. They tend to be active to some extent in any language, especially from the standpoint of diachrony. For example, Romance languages do not have consonant gradation in synchrony, but weakening and strengthening have been active in historical stages, from Latin to French (most prone to phonological weakening) or from Latin to Italian (where strengthening conditions inherited from Latin are still obvious through geminate consonants in the Standard dialect and Southern Italo-Romance dialect, which have even developed new mechanisms of consonant strengthening, such as the so-called *radoppiamento sintattico*).

Greenbergian indices, as shown later in Table 2.1 in Section 2.5, are a reduction of more complex reality, made available through a survey of dialect networks. For instance, a language such as Tseltal can be considered as fairly agglutinative and synthetic, with a touch of polysynthesis, as prefixation is crucial in many respects in its grammar (for possession and ergative marking).

Yet, if we apply the same grid as in Table 2.1, some dialects may differ, increasing the rate of synthetic patterns or, on the contrary, analytic patterns. To some extent, results as shown in Table 3.12 in Section 3.7 are conditioned by this kind of fine-grained variation of typological properties within the dialect network.

Cedergren's Model

As pointed out by Blevins (2004),

> '... sound changes fall into many different categories. It is noteworthy that the majority of commonly attested sound changes in the world's languages are mirrored by synchronic alternations of precisely the same type. Sound changes give rise to changes in segment inventories, segment sequences, syllable types, stress patterns, tone patterns, and feature distribution.'

As we will soon see, not only should all these factors be taken into account in the survey of any sound change in progress, for any of the parameters enumerated earlier, but they should, or could, also be hierarchized, according to local conditions (the typological properties of each language). Of course, many other processes can be observed in the phonological evolution of natural languages, but the basic list in 'Main types of sound change' accounts for most of them: for example, primary diphthongization, as in Latin *festa* > Spanish *fiesta*, 'feast, party' can be considered a subtype of vowel dissimilation (explained earlier), as well as secondary diphthongization, as in Latin *nocte* > Portuguese *noite*, 'night', results from the lenition of an implosive stop into a glide.

An outstanding model for computing structural embedding of variation (and thereof sound change) has been provided by Cedergren in her dissertation on sociolinguistic variation in Panama for the aspiration of implosive *s* in Panama Spanish (Cedergren, 1973; Ramírez, 1996). She introduced the following formula to take into account contextual factors such as a range of probability to evaluate the variation of what matches the lenition parameter. In this case, the parameter at stake is a kind of lenition (the eighth phonological process in the list, namely $s > h$ for a so-called implosive sibilant; that is, syllable or word-final). Cedergren's model of probability coefficients for Panama Spanish's implosive sibilant aspiration ($s > h$) is defined by the following change probability:

$$(1 - p) = (1 - p_0) \times (1 - p_a) \times (1 - p_b) \times (1 - p_c) \times (1 - p_d) \times (1 - p_e),$$

where p stands for the probability for a variant to appear in a specific context, p_0 for the mean value over all other contexts, whereas p_a, p_b, p_c, p_d, p_e match the effects induced by each contextual factor, as enumerated here:

a : lexical and syntactic category (adjective, determinant, noun)
b : grammatical function (monomorphemic ⟨más⟩, 'more'; number (plural) ⟨los muchachos⟩, 'the boys'; verbal inflection (subject agreement) ⟨tienes⟩, 'you have').
c : position within the word (internal ⟨estar⟩, 'to stand, be (somewhere)'; or final ⟨cantas⟩, 'you sing').
d : phonotactic context (consonant, vowel, or pause: _C, V_V, _#)
e : speech style (formal vs. informal)

The following are the results for implosive *s* aspiration in Panama City $p_0 = 0.21$.
Compare this with the following figures for the probability that the phenomenon would positively occur in speech, from Cedergren's sample:
Adjective $p = 0.66$, noun $p = 0.58$, determinant $p = 0$;
Monomorphemic $p = 0.49$, plural $p = 0.08$, verbal inflection $p = 0$;
internal position within the word $p = 0.62$, final position within the word $p = 0$;
preconsonantic position $p = 0.89$, intervocalic position $p = 0.49$, before pause $p = 0$;
informal speech $p = 0.15$, formal speech $p = 0$.
These figures actually point to what could be called a grammar of sibilant aspiration in Panama Spanish. Sound change in progress is by no means randomly distributed as the $p = 0$ paradigms suggest, as opposed to the variably positive paradigms in competition within this emerging variable grammar of aspiration. The various p values assigned to each lexical and syntactic pattern or phonotactic (that is, contextual) settings or speech style (that is, registers) have a quantum effect on the structural rule, which is explained here.

Rule of sibilant aspiration in Panama Spanish (Cedergren, 1973)

The rule is expressed as follows:

$$S \rightarrow \langle \underset{R}{H} \rangle / \_\_\_]\#\# \langle C|V|P \rangle \qquad (2.1)$$

where] stands for closed syllable boundary, #: word limit, C: Consonant, V: Vowel, P: Pause, R: Register (formal/informal).

The interplay of segment inventories (p_i), stress patterns (p_s), tone patterns (p_t), and feature distribution (p_f), as quoted by Blevins (2004), also causes interferance more or less inherently or according to each language or dialect, so that formulas can be complexified to include segment sequences syllable types (p_t) and sound sequences (p_q), which were partially taken into account previously as (p_b) and (p_d). Such an intricate set of patterns also includes, for example tone patterns, in addition to stress patterns, which may prove useful when analyzing variation in an Ottomanguean language (for example, Mazatec), or an Indo-European language with pitch stress, as Lithuanian (in this case, p_a should be added, taking into account inflectional classes for each word class). Moreover, as pointed out before, each probability condition (or broadly speaking, context) should, or at least, could be hierarchized and weighed, according to the specific typology of the language. For instance, Cedergren's p_b variable was somewhat heterogeneous, including syllable type features (such as monosyllabicity), mingled with inflectional criteria (plural number for nouns and verbal inflection). All these varied, though consistent trends of distribution conditioning the phenomenon under scrutiny amount to what linguists call morphophonemic p_b or morphophonological patterns. They amount to a variable state grammar, or a quantum socio-phonetic grammar, for the phenomenon, which combines with other morphophonemic variables of the same kind in the language.

Further complexification of probability quanta for phonological variation

What Cedergen's pioneer research encompasses, and what has paved the way for the notion of 'variable rule', has to do with the holographic distribution of variation within the web of the lexicon, grammar, and even discourse and speech situations in social settings. Since the first quantitative surveys of sociolinguistic variation by William Labov in the 1960s[7] (Labov, 1970, 2006, 1972b,a), the paradigm of variation studies has become central in the study of the interaction between language and society, especially from the standpoint of social stratification and social networks.[8]

Sound correspondences in the world's languages have been comprehensively listed in the recent survey by Brown et al. (2013) of the ASJP database,[9] amounting to no less than 692 pairs from 5,117 languages and dialects pertaining to 443 genera and 138 families, or linguistic stocks. Correspondence pairs for consonants (such as l:r) are put together according to the main

---

[7] More information on this central paradigm in quantitative sociolinguistics is available on the author's Web pages: www.ling.upenn.edu/w̃labov/.

[8] See also Milroy's pioneering work in the specific domain of sociolinguistic networks, of great relevance to complexity theory (Milroy, 1980; Milroy and Milroy, 1978).

[9] Automated Similarity Judgment Program, available online at http://asjp.clld.org/.

relevant features (for example, laterals vs. rhotics), the number of genera (or phylogenetic clusters, such as Germanic or Slavic, within the Indo-European clade) involved, and the mean ranking of these sets within the genera concerned, pointing at the relative density of the phenomenon within local phylogenetic domains. These are but a few phenomena parsed according to basic criteria such as frontness, backness, sonority scale (sonority, rhoticity) and the opposition between sound gestures (relying on phonation, as laryngeals, or on articulatory constriction in the oral cavity). Of course, the 692 pairs involve more criteria, but this sample will be sufficient here in order to highlight the taxonomy: in the same way that lexicon and grammar can be described as bounded systems, so can sets of sound changes and correspondences in the world's languages. An interesting property of the pairs listed here is that the actual density of the sets varies according to sub-criteria; for example, consider articulatory place (coronal vs. dorsal) or sonority (voiced vs. unvoiced): compare the density per genus of the labial with the coronal voiced aspirated stops (bh:b vs. dh:d), that is, 41.67 vs. 30.77%, or the g:k pair (12.86%) versus the kh:k pair (17.25%), whereas other types, such as palatals vs. nonpalatals are more consistent (around 15%).

A sample of sound correspondences from the ASJP database (Brown et al., 2013)

- Frontness
  $t^y : t$ (unvoiced palatal vs. nonpalatal coronal stops): 5; 14.71%
  $n^y : n$ (palatal vs. nonpalatal coronal nasal sonorants): 4; 14.29%

- Backness
  $\chi : x$ (unvoiced uvular vs. dorsal fricative): 10; 23.26%
  $q : x$ (unvoiced uvular stop vs. dorsal fricative) : 7; 11.86%

- Sonority
  $g : k$ (voiced vs. unvoiced dorsal stops): 31; 12.86%
  $d : t$ (voiced vs. unvoiced coronal stops): 32; 12.08%

- Rhoticity
  $l : r$ (lateral vs. rhotics) : 55 genera; 20.68%
  $r : d$ (rhotic vs. voiced coronal stops): 23; 9.31%

- Phonation
  $b^h : b$ (voiced aspirated labial stop vs. non aspirated labial stop): 5; 41.67%
  $d^h : d$ (voiced aspirated vs. nonaspirated coronal stops): 4; 30.77%
  $k^h : k$ (unvoiced aspirated vs. nonaspirated dorsal stops): 17; 25%
  $h : x$ (laryngeal vs. dorsal fricatives): 27; 20.45%

2.4 Complex Networks

It is now necessary to digress from our main topic and introduce some notions of complex networks, a topic which is relevant to many problems of language dynamics considered in this book. In this section, we present a brief overview of a few basic concepts, models, and terminology of complex networks theory. If space represents a basic concept on which science is founded, it cannot be different for complex networks, which are actually ways to represent the underlying geometry of a complex system.

2.4.1 General considerations about complex networks

When the effect of the spatial structure of a system can be neglected and a process can be described as approximately homogeneous in space, we can call the system a '0-dimensional system' to underline the lack of space dependence: for example, describing a biological population in ecology only requires determination its time-dependent population size $n(t)$. On the other hand, in other models, the dependence on one or more spatial coordinates is relevant: a motion along a line can be described as a '1-dimensional' motion; a '2-dimensional model' is employed for describing a systems moving on a plane or a surface, and '3-dimensional models' are used for processes taking place in space. However, in many situations, the underlying space is of a very different type.

In complex systems, made up of interacting units, a network is the natural and general way to represent the structure of the system. A *network* (called *graph* in mathematical terminology) is composed of *nodes* (also called *vertices*), usually representing the units of the system, with pairs of nodes connected by *links* (or *edges*); a link represents the interaction between two nodes.

Networks can be regular networks, in which each node has the same *degree*, that is, the same number of links. A chain is a particularly simple type of regular lattice, in which a set of N nodes $\{i\}$, with $i = 1, \ldots, N$, are connected to each other consecutively, so that each node i has degree $k_i = 2$; thus, each node has only the two neighbors $i-1$ and $i+1$. A one-dimensional chain approximates a continuous (one-dimensional) space, if the lattice constant, the distance between two neighbors at which nodes are assumed to be located, is small enough; similar considerations hold, for example, for two- and three-dimensional lattices. Thus, a continuous space can be seen as a particular limit of a suitable regular network.

However, a network can also describe more complex types of space for which there is in general no continuous analog; this is because it can connect a set of nodes in any arbitrary way. Such a general framework well describes networks of interactions, such as that between neurons, whose connections are long-ranged and entangled; the professional or personal relationships between individuals, as revealed by, for example, mail activity or collaborations; the internet of all connected computer and servers, as well as the WWW of all the web sites connected to each other through hyperlinks; communication networks, such as road networks or the network formed by airports and the airplane routes connecting them to each other; biological processes as well as the relations between biological species in a food web; economic relations between, for example, companies; etc.

In particular, there are various examples of how complex networks enter the study of language.

- *Network of individual interactions.* In interactions between speakers within a linguistic group, it is possible that communications do not take place uniformly but along special links defined by family bonds, between colleagues or more in general across the same working environments, at a social level in the same social group or within the same ethnic community. If an individual-based model is used to describe these interactions, the underlying structure may not appear clearly; it is best described in terms of a network of interactions, which defines the internal connections and the otherwise invisible external boundary of a group. Complex networks are needed to characterize social interactions, since the underlying social structure is typically highly heterogeneous among speakers. In a group of N individuals, one can imagine each individual i sitting at a node of a network, connected by links to other individuals. Given a pair of individuals i and j, one can associate a number w_{ij} expressing, for example, the rate of communication. The values w_{ij} can be organized into an $N \times N$ matrix called the *adjacency matrix*, which specifies the way in which the units interact with each other. An element $w_{ij} = 0$ would mean there is no interaction between i and j individuals. The *adjacency matrix* quantifies not only the strength but also the direction of links between individuals, if it is not symmetrical ($w_{ij} \neq w_{ji}$).
- *Networks of linguistic similarity.* One can also consider a given linguistic feature (or a set of them) in different languages as units of the system and use such feature(s) to study the relationships (for example, the similarity) between the various languages through a network-based approach. Various examples will be provided in the following sections using case studies, in particular, the network of morphological features determined by the syntax rules used in different languages; the network defined by the different phonological features; or the lexical network.
- *Networks of mutual intelligibility.* A language family, whose languages can be considered as nodes, can also be studied by looking at their mutual intelligibility, which is quantified through an adjacency matrix. The elements of the matrix represent the scores obtained in suitably designed tests involving speakers of the various languages.
- *Semantic network.* An example of a semantic network is obtained by choosing the nodes of the network as words used in a certain text and connecting pairs of them if they represent two words appearing in the text consecutively. A weight can be assigned to each link, determined by the rate of occurrence of the corresponding two-word sequence. One can define a similar network in other ways, for instance, looking at how many hyperlinks connect the different pages of online encyclopedias, such as Wikipedia.

Before moving on to more technical notions about networks, a few remarks are in order.

First, although we approximate networks to be static, it is to be noticed that a network-based representation of a complex system has relevant consequences on the dynamic side. Even a static network (fixed in time) may have time-dependent processes which can take place across the network such as information spreading or opinion diffusion, that strongly depend on the type of network.

Furthermore, a network may be intrinsically time-dependent, since in real networks, links can be created, switched on or off, or simply rewired.

In addition, we remark that although a network-based framework may effectively capture many relevant complex features of a system, in some cases, for example, when the relation between more than two languages at time has to be considered, the system may be better represented within more general frameworks, such as historical glottometry, discussed in Chapter 4.

Finally, since the information about a (linguistic) system can be often provided by different types of networks applied to the same system, the full characterization of the structure of a complex system is possible only by the concept of *multiplex*, meaning a system of overlapped and interacting networks; a well known example is the multilayer nature of the social network.

2.4.2 Features of a complex network

In this section, we summarize a few quantities that are helpful for specifying and classifying the type of complex network.

- A network with N nodes is defined by the associated $N \times N$ adjacency matrix $W = \{w_{ij}\}$ defining the interactions between the nodes ($i, j = 1, \ldots, N$). The matrix W is symmetrical ($w_{ij} = w_{ji}$) if the network is undirected (symmetrical links), while in the case of a directed network, it is asymmetrical ($w_{ij} \neq w_{ji}$). The elements of the adjacency matrix measure the strength of the node–node interactions or flow between nodes. When only the presence or absence of a link is of interest, one considers non-weighted networks, characterized by an adjacency matrix W with elements that are either $w_{ij} = 0$ (signaling absence of interaction) or $w_{ij} = 1$ (presence of a link).

- The degree (or connectivity) k_i of a node i is the number of links of the node (for directed networks, there is an in-degree $k_i^{(in)}$ and an out-degree $k_i^{(out)}$). A highly connected node is often called a hub; usually hubs are present in strongly heterogeneous (complex) networks. The degree distribution $P(k)$ of an undirected network (as well as the corresponding in- and out-distributions $P^{(in)}(k)$ and $P^{(out)}(k)$ of a directed network) is an important quantity defining the type of network and gives the probability that a given node has exactly k links. For example, scale-free networks can be recognized from their power law degree distribution, $P(k) \sim 1/k^\gamma$ for $k \gg 1$, where γ is the degree exponent. For undirected networks with N nodes and L links, one can define the average degree $\langle k \rangle = 2L/N$ (Figure 2.1).

- The average path length ℓ associated with a network is obtained by averaging the shortest paths between all pairs of nodes; this length provides an estimate of the overall navigability of the network. Many complex networks, despite their large size, have relatively short paths between pairs of nodes, as in the case of *small-world networks*.

- The clustering coefficient of a node i is defined as

$$C_i(k_i) = \frac{2E_i}{k_i(k_i-1)}, \qquad (2.2)$$

where E_i is the number of links connecting the k_i neighbors of the ith node to each other (see for example Barabási and Oltvai [2004]; Albert and Barabasi [2002]). The tendency to form clusters in a given network is measured by the average clustering coefficient C.

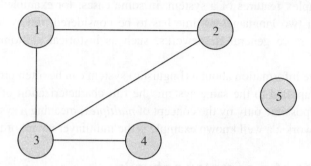

Figure 2.1 An example of an undirected graph with $N = 5$ nodes and $L = 4$ edges. The connectivity of the nodes are $k_1 = 2, k_2 = 2, k_3 = 3, k_4 = 1, k_5 = 0$; the average degree is $\langle k \rangle = \sum_i k_i / N = 8/5 = 1.6$. Only nodes 1, 2, and 3 have a non-zero clustering coefficient, $C_1 = C_2 = C_3 = 1; C_4 = C_5 = 0$.

In general, the quantities $\langle k \rangle$, ℓ, and C measure the structural properties of complex networks and clearly depend on the numbers N of nodes and L of links. On the other hand, the degree distribution $P(k)$ is independent of a network's size and captures some generic features of the network (Albert and Barabasi, 2002).

The concept of a (connected) subgraph can be useful. A graph G_1 is a subgraph of a graph G if all its nodes and links also belong to G. We can find trees, cycles, and complete graphs among the possible subgraphs of a given network.

Networks can contain highly connected nodes (hubs). The complex character of a network is related to its heterogeneity and therefore to the diversified characteristics of its nodes.

2.4.3 Types of complex networks

The main types of networks are as follows.
- *Connected networks.* A connected network is one where, given two nodes i and j, there is at least one path $P(j|i)$ connecting i and j. In other words, there are no isolated nodes or sub-networks.
- *Fully connected networks.* A network that has a set of nodes which can all communicate with each other. Sometimes, the presence of such a network is not even mentioned, but its existence should be kept in mind. Fully connected networks are also called *complete graphs* or *all-to-all connected networks*.
- *Regular lattices.* The one-dimensional lattice (chain) mentioned earlier is the simplest example of a regular lattice. In general, in this type of network, the structure of a basic group of nodes is repeated periodically. The generalization of a chain to a 2-, 3-, or N-dimensional lattice is straightforward. These types of networks are relevant in the representation of surfaces and crystals,

- *Regular networks.* In a regular network, each node has the same degree. It is necessary to point out that regular networks may have peculiar structures, very different from that of the regular lattice.
- *Random graphs.* There are different but equivalent versions of the Erdős–Rényi random graph, depending the way in which they can be constructed (see Albert and Barabasi [2002]; Erdős and Rényi [1959, 1960]).

In one of these versions, the links in an initially disconnected set of nodes are created with a given constant probability p_0 considering all possible links, so that eventually the expected average number of links is $\langle L \rangle = p_0 N(N-1)/2$. The corresponding clustering coefficient is $C(k) = \langle k \rangle / N = p_0$, which is independent of the degree of the node. The mean path length is proportional to the logarithm of the network size, that is, $\ell \sim \log N$, as in a small-world network (explained a little later). Figure 2.2 shows a realization of a random graph for $N = 6$ nodes and $p_0 = 0.7$. In Erdős–Rényi random graphs, by construction, each node has eventually a number of links close to $\langle k \rangle$. It is possible to show that for large N, the graphs have a Poissonian degree distribution,

$$P(k) \simeq e^{-p_0 N} \frac{(p_0 N)^k}{k!} = e^{-\langle k \rangle} \frac{\langle k \rangle^k}{k!}, \qquad (2.3)$$

which implies an exponentially decreasing k tail. In turn, this means that in random graphs, the node connectivities rarely deviate from $\langle k \rangle$. Thus, Erdős–Rényi random networks can be basically considered or effectively behave as homogeneous networks.

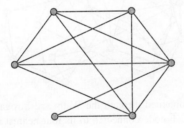

Figure 2.2 A realization of the Erdős–Rényi random graph with $N = 6$ nodes and probability $p_0 = 0.7$. The total number of edges expected is $\langle L \rangle = p_0 N(N-1)/2 = 10.5$; notice that for $p_0 = 1$, a fully connected network is expected.

- *Small-world networks.* In many biological and social networks, one finds that the connection topology is neither completely regular nor completely random. This feature is also observed in the neural networks of a worm, the *Caenorhabditis elegans* (Albert and Barabasi, 2002). Based on the observations of Albert and Barabasi, Watts and Strogatz proposed the *small-world network model* which can interpolate between regular and random networks (Watts and Strogatz, 1998) as a varied parameter. Such a network is referred to as a 'small-world' network because it is highly clustered, like a regular lattice; however, at the same time, it shows small path lengths, a feature common in random graphs (Watts and Strogatz, 1998). In fact, small-world networks can be considered a superposition of regular lattices and random graphs (Dorogovtsev and Mendes, 2003). The basic algorithm of Watts and Strogatz to construct a small-world network can be summarized as follows (Albert and Barabasi, 2002).

(i) Start from a regular ring lattice, in which there are N nodes, each node being connected to its neighbors through k edges.

(ii) Randomly rewire each edge with a given probability p_1.

Figure 2.3 shows the realization of a small-world network generated using a rewiring probability $p_1 = 0.4$ in a regular ring with $N = 24$ nodes, each connected to its four nearest neighbors ($k = 4$). The general conditions for the applicability of this algorithm is that $N \gg k \gg \ln(L) \gg 1$, where the intermediate inequalities guarantee that the random graph is connected (Watts and Strogatz, 1998; Albert and Barabasi, 2002). The parameter of the network regulating its randomness character is the rewiring probability p_1 (randomness increases with p_1): for small $p_1 \to 0$, the clustering coefficient $C(p_1)$ and the average path length $l(p_1)$ converge to those of a regular lattice (of a random graph). For a large interval of probabilities, the networks formed are at the same time highly clustered (large C as in regular lattices) and have ℓ comparable to those of random graphs (Watts and Strogatz, 1998). The degree distribution of the networks behaves similarly to those of random graphs, with an exponential tail; moreover, in this type of networks the degree does not fluctuate much (Albert and Barabasi, 2002).

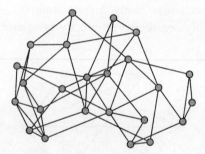

Figure 2.3 Small-world network obtained through the Watts and Strogatz's procedure from a regular ring with $N = 24$ nodes; each node connected to its four nearest neighbors ($k = 4$) using a rewiring probability $p_1 = 0.4$. Note that duplicate edges were not allowed.

- *Scale-free networks.* Many real large networks are characterized by a power law degree distribution $P(k)$. Due to the scaling invariance of the power law function, such networks are referred to as *scale-free networks*. The complex network models discussed till now cannot reproduce this feature. Scale-free networks have a strongly heterogeneous topology, with more low-degree nodes than nodes with high degree.

Albert and Barabási realized that in scale-free graphs, the network dynamics plays a crucial role in forming real networks and suggested that there are two basic mechanisms underlying the origin of scale-free networks. First, real large networks grow by continuous addition of new nodes. Second, in random graphs and small-world networks, the connectivity probability between nodes is independent of the nodes' degree by definition so that there is no preferential attachment. On the basis of these observations, the Albert–Barabási model of scale-free networks was introduced; it is basically a growing network model (Albert and Barabasi, 2002), construction of which can be performed using the following algorithm.

(i) Start from a network with a given number of nodes $\tilde{N}$.
(ii) Add a new node n and link it to another node i already in the network through a preferential attachment, that is, by selecting randomly the node i among the other nodes with a probability

$$\Pi_i = \frac{k_i}{\sum_j k_j}, \qquad (2.4)$$

proportional to the degree of the node i.
(iii) Repeat this process of linking the same new node n to other nodes in the network, until node n has m edges.
(iv) Reiterate this procedure restarting from (ii) until the network has reached the desired size N.

Figure 2.4 shows a small scale-free network of $N = 20$ nodes, constructed with a preferential attachment criteria using $m = 7$ as parameter, according to the Albert and Barabási algorithm. It is possible to see that even in this small network, a few hubs have appeared, characterized by a higher number of links with respect to the other nodes.

When constructing larger networks, this algorithm leads to a power law degree distribution $P(k) = 1/k^\gamma$, with degree exponent $\gamma = 3$; the average path length ℓ is smaller than in random graphs for any value of N (Albert and Barabasi, 2002); moreover, the clustering coefficient follows a power law $C = N^{-3/4}$.

- *Tree.* In linguistics, in particular in cladistics and distance-based data analysis methods, tree networks are of particular relevance. A tree network is an undirected graph in which,

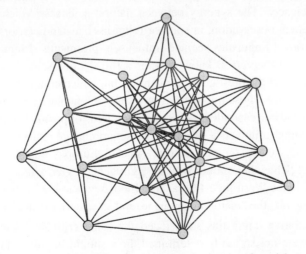

Figure 2.4 Scale-free network with $N = 20$ nodes constructed using a preferential attachment algorithm with $m = 7$. Note that some hubs are already clearly visible, despite the small size of the graph.

for any two vertices i and j, there is only one path $P(j|i)$ connecting i to j; there are no cycles. Thus, a tree with N nodes has $N - 1$ links.

- *Spanning tree.* A spanning tree is defined as a sub-network of a given undirected (and connected) network, constructed as a tree including all the vertices of the network, with a minimum possible number of links; in general, a network can have several spanning trees. As for a general tree, a spanning tree of a network with N nodes also has $N - 1$ links.

This completes our partial overview of networks; other concepts related to (complex) networks will be introduced along the way, as they are needed.

2.5 Trees vs. Networks

As pointed out earlier, language is by no means a living organism, and although it has been heuristically compared with living organisms in the past, one should always remember that this is but a metaphor. Even August Schleicher, famous for his enthusiasm for the integrative potential opened up by the evolutionary theory proposed by Charles Darwin, a skilled horticulturist himself, and interested in the selective power of plant domestication, always made it clear that he intended organicity as a metaphor. As Tort pointed out in a remarkable essay (Tort et al., 1980), in the late nineteen century, Darwinism provided an outstanding opportunity for the emerging paradigm of comparative linguistics to complete its fusion with natural sciences. The successful comparison of geographically distant languages and genera, and the availability of intermediate dialect tokens, allowing the identification and classification of the inner structure of linguistic 'families' as 'species' through Stammbaums (phylogenetic trees) had paved the way for biology to make important steps forward using Lamarckism; it has also helped to develop more dynamic taxonomies, whose complexity can be explained by large-scale diffusion processes. As time went, dynamism won over fixism and conservation science. The synergy between natural sciences, which gained legitimacy over centuries of patient observation, in spite of religious bans and censorship, and the brand new method of historical linguistics seemed immensely promising. Linguistics soon entered the field of positive and theoretical knowledge as a full-fledged science, over the traditional philology, which inherited the prestige of humanities, but which could not claim any modeling power. It became a science whose foundations were laid on phonetic laws (see Collinge [1985] for a critical survey) and correspondences between hundreds of positive cognates in the lexicon and grammar, according to universal processes of sound change and grammatical robustness within linguistic stocks over continents, from Asia to Europe (that is, the Indo-European 'language family', at Schleicher's time, with its eight genera: Germanic, Slavic, Italic, Celtic, Albanian and Greek in the West, Indo-Iranian and Indo-Aryan in the East, before the discovery of the central missing link, the Anatolian language, and an eastern extinct language, Tocharian).

However, comparative linguistics soon faced the same criticism as evolutionism did in biology: if genera were not so easily determined, how should linguistic families be defined? On what criteria? How should they be delimited? Can linguistic stocks be lumped into superordinates, such as phyla? How can linguistic stocks or even phyla pass through the test of cross-fertilization: a phenomenon even more obvious in the realm of languages? This raised, among many other questions, the issue of types over genera.

As Tort puts it (Tort et al., 1980, p. 20), linguistics had to face the same cascade of questions as emerging Darwinism: how to account for the Malthusian principle of geometric expansion, the disappearance of ancient forms, differentiation processes, genealogical ramification, diverging vs. converging traits, the blossoming of innovative forms from the pool of ancient ones, hierarchy of clusters and organizing patterns, the fading, the bursting and the decay of intermediate stages, extinction, survival and resilience of higher level organisms. Schleicher's Stammbaum was challenged by the prerequisites of modern taxonomy; whereas the new paradigm of comparative linguistics could not rely on an already elaborated method to establish nomenclature for biological objects, as provided by naturalist, Georges-Louis Leclerc de Buffon (1707–1788) or Carl von Linné (1707–1778). General linguistics was still to be founded; an endeavor to which such figures as Wilhelm von Humboldt (1767–1835) and Hugo Schuchardt (1842–1927) contributed, before the masterly breakthrough by Ferdinand de Saussure (1857–1913), but these were still individual contributions: neither a general theory nor a unified method. Humboldt can be considered as a precursor and Schuchardt as an opponent to the Neogrammarian doctrine. Saussure succeeded in providing a cohesive theory out of the Neogrammarian paradigm, but his *Cours de linguistique générale*, published at the eve of World War I, was not a major handbook, but rather the compilation of his lectures by two students, who sensed the innovative character of his thoughts. Of course, leading figures in Neogrammarian theory, such as Hermann Paul, or Karl Brugmann and Hermann Osthoff, did produce treatises and major essays, directed at the synthesis of a comparative method, especially with the aim to strengthen the legitimacy and the hegemonic status of the Neogrammarian school (Paul 1879, 1880; Brugmann and Osthoff 1880), which shared many principles and concepts with contemporary general linguistics and even modern theories in this field (see Scheer and Ségéral [2016]). However, these were somewhat dispersed attempts, which did not work out into anything comparable to Charles Darwin's three seminal contributions to evolutionism, a subparadigm within biology (Darwin, 2007, 1868, 2004). Moreover, the whole architecture of nineteenth century comparative linguistics was challenged by criticism on several issues, such as Johannes Schmidt's (1843–1901) wave model (Wellentheorie), which viewed the separation of Indo-European languages as diffusion of waves created when a stone is thrown into a body of water (Schmidt, 2017), disputed the sound laws introduced by the Neogrammarian model. Nowadays, Schmidt's model is described as a kind of reaction–diffusion model. This approach was later to be used considerably in dialectology, through the works of Matteo Bartoli (1873–1946), in a paradigm strongly rooted in Romance dialectology, especially in Italy, called Neolinguistica: an emerging paradigm whose name itself challenged a posteriori the Neogrammarian formerly hegemonic model.

Some opponents of late nineteenth century comparatism were even more pugnacious and restless than diffusionist thinkers, quietly challenging trees and classifications, used beyond the geographical range. One element which causes serious dilemma in current theories of linguistic change and structures based on gradual diversification of patterns inherited from more or less prestigious ancestors, such as Vedic Sanskrit or, later, Proto-Indo-European, is the mixed and intricate nature of creole languages, as discovered by Hugo Schuchardt (see Glenn Gilbert's recent reader in Schuchardt [1980]). This unexpected empirical field of massive and drastic change in all areas of phonology, grammar, and the lexicon pointed implicitly at four kinds of variables which could account for such phenomena as the genesis of creoles and pidgins.

(i) Some yet unknown universal principles (harbinger to what will later be called 'universal grammar' by Noam Chomsky) of superordinate evolutionary laws making the sudden readjustment of linguistic systems possible.
(ii) More or less uniform typological traits shared by all these new languages over any phylogenetic determinism from the donor languages (French, English, Portuguese, Spanish, and Dutch).
(iii) Contact incidence (through the donor language, but also through adstrata and in some cases, substrata) and areal considerations.
(iv) Phylogenetic resilience of the donor languages.[10]

Creoles and pidgins also opened a whole range of questions for evolutionary linguistics which are still the focus of much discussion, as well as the so-called bioprogram, suggesting that the emergence of these linguistic systems could be compared to a kind of natural simulation of self-organization by the language itself, mirroring basic mental operations and fundamental categories involved in the emergence of language as an attribute of the human species (Bickerton, 1981, 1992). Another much debated issue had to do with the strong positivist trend of the Neogrammarian school of comparative linguistics, which was fueled by discrete empirical data, clear-cut paradigms, and univocal rules, and which could hardly cope with the inner variation and polymorphism overwhelmingly present in dialectological data. Although Neogrammarians were not as interested by evolutionary issues as Schleicher and the first comparatists, this was a serious challenge from the evolutionary standpoint, as Darwinism implied that it was from variation pools of individuals, varieties, or species, that the fittest variants emerge as survivors and new challengers. Nevertheless, Neogrammarians were convinced that dialectology was a 'new frontier' for their method, and their descriptive models, based on a prototype of what would later be called 'complementary distribution', was perfectly fit for this descriptive task. Needless to say, dialectological, and especially geolinguistic data presented several phenomena that challenged their models: why did Latin *apis* ('bee'), expected to become *év* in Modern French according to standard rules of evolutions, have to compete with so many different phonological reflexes and different lexical options throughout the Gallo-Romance territory (in short, the territories where French is spoken nowadays in Europe as an official language), such as *mouche à miel, guêpe* (out of a confusion with Latin *vespa* 'wasp'), and have so many queer phonological reflexes (*ép, és, wés, é,* etc.), often unexpected or even irregular, such as *abeille* in Modern French, a loan from the so-called old Occitan varieties or old Provençal? According to Jules Gilliéron (1854–1926), this is typically the kind of question which cannot be answered by scholarly etymology. One has to

---

[10] Kihm (2006) suggested a fourfold model to handle the empirical and epistemological complexity of creoles in terms of networks or bundles of structural traits: Universal grammar's networks, typological networks, areal networks, and phylogenetical networks. Although to search for overt reference to such an elaborated framework in Hugo Schuchardt's works would be an obvious anachronism, one can say that these questions implicitly came forth out of the data he produced and in the way he contradicted the Neogrammarian doctrine on the basis of phenomena observed through the initial survey of creoles and pidgins.

resort to folk etymology, that is, therapeutic strategies to frame oppositions in the lexicon. This meant that psychology had to be introduced in the study of language in a far more creative and less mechanic way than the Comparatists of his time would ever accept through predictable etymology. Gilliéron was not content with just placing a few arguments on the table, and pontificating upon them: together with a unique and astonishingly good descriptivist, though a perfect layman in linguistics, Edmond Edmont, he managed in the short span of four years (1897–1901) to collect, and publish in eight years (1902–1910) a monumental linguistic atlas of France, ALF (1421 maps), covering 639 localities (Jules and Edmont, 1902–10) (Brun-Trigaud et al., 2005), full of first-hand data, which overshadowed all previous empirical compilation ever done by comparatists on any single language or genera (Gallo-Romance, as a branch of Romance languages, matching typically the concept of sub-genera within the Indo-European family). This new set of data also questioned the strategic issue of continuity vs. discontinuity and discreteness of such basic categories as language and dialect. Can dialect boundaries be traced, out of bundles of isoglosses? Where does a dialect start, where does it end? If the question of differentiating objectively between languages and dialects was by itself a hard nut to crack, geolinguistics made things even worse, as the impressive quantity of fresh data from hundreds of localities showed intricate patterns of overlapping phonological systems and grammars, and boundless lexicons, as evaluative factors from folk linguistics could constantly reshuffle the cards and, consequently, shuffle areas on the maps. Conditions of predictability were also strongly challenged. To solve in all these issues, complexity theory is welcome and can likely provide many solutions, as claimed in the present volume.

Gigantic geolinguistic survey outcomes such as Gilliéron's ALF, puzzling emerging and self-organized languages such as creoles and pidgins as studied by Hugo Schuchardt, were not the only empirical and epistemological hurdles that comparative linguistics and the tenets of evolutionary linguistics faced; although, previously they were considered important issues in the emergent paradigm of early comparativism, or during the transition between the Schlegelian or Sleischerian and Humboldtian tradition, and the Neogrammarian epoch. An aporia still haunted the whole edifice of evolutionary linguistics: typology and the quest for reliable taxonomic classes.

With the contributions by the Schlegel brothers, the next set of morphological types to be recognized was the opposition between three main types: isolating (root words, without inflection), agglutinating (mechanically concatenative), and inflectional (organically altered or modified), as in Schleicher's 'algebra':

Isolating → R,
Agglutinative → Rs,
Inflectional → RxSx,

where R stands for a single root, without any inflectional alternation, as in Sinitic (Chinese, Cantonese) or Austro-Asiatic (Vietnamese, Hmong) languages; Rs for a root + suffixes, as in Altaïc languages (Turkic, Mongolian, Tungusic) or Uralic languages (Hungarian, Finnish, Samoyed); RxSx for languages modifying roots into arrays of variable stems + suffixes, as Romance languages and most Indo-European languages.

However, as soon as linguists began to explore linguistic families other than Indo-European and Semitic languages, they soon discovered that most linguistic families show somewhat or even highly heterogeneous patterns. Within the Uralic linguistic stock, for instance, languages like Estonian or Livonian happen to rely intensively on inflectional patterns, as compared to Vepsian, in the same small *local genus* (Finnic). Standard Finnish indeed looks like a fairly agglutinative language at first sight, although stems vary according to syllable structure (compare *talo*, 'house': nominative singular vs. *talo-n*, 'house': genitive/accusative, uninflected and merely agglutinative, to *silta*, 'bridge': nominative singular vs. silla-n, 'bridge': genitive/accusative singular, with progressive assimilation of *-lt-* → *-ll-*: a so-called juncture process, triggering stem alternation), but many dialects, such as the Turku and Rauma varieties in the southwest, or southeastern dialects from the Isthmus of Carelia, display impressive sets of alternating stems out of lenition processes, which in general delete many consonants within stems as in Estonian, Votic, and Livonian. The inner heterogeneity of gigantic linguistic stocks, such as the Niger–Congo languages, is also striking: compare the highly agglutinative Bantu languages with Kru or Gur languages, which show respectively strong isolating or inflectional trends (Hyman, 2011). The same could be said for intricate groups such as the Oto-Manguean languages, with Chinantec and Chatino matching the isolating type, whereas Zapotec and Mazatec are highly inflectional: although all are highly tonal languages, like the Sinitic and Austro-asiatic languages. In the case of Mazatec and Mixtec, as Pike pointed out (Pike, 1948) when first describing their morphology for various purposes, the lexicon of these languages seem to work as a hologram, with intricate patterns of compounding (and therefore, of isolating subsets of uninflected roots) and alternating stems, using a tricky process which blurs the whole game: stem suppletion (as in French *je vais*, 'I go', but *j'irai*, 'I'll go' vs. *j'allais*, 'I used to go', with suppletive stems *v-*, *ir-*, *al-*, from three intertwined different verbs in Latin: *vadere*, 'to cross (one's way)', *ire*, 'to go', *allare*, 'to stalk'). The phenomenon is so striking and pervasive in these languages, that Pike worked out a whole descriptive model within structural/functional linguistics called 'tagmemics' to account for it. According to tagmemics, local combinatorics at all levels of linguistic systems (phoneme, syllable, word, stems, phrases, sentences, paragraph, and discourse) make up languages and linguistic types, through matrices, described by a specific algebra. Tagmemics comes close to the more recent paradigms emerging in phonology and morphology in modern theoretical and descriptive linguistics (Bybee and Hopper, 2001).

Last, but not least, one can say that many languages display evidence of heterogeneity in this respect: for example, English can be considered as predominantly isolating, yet it has a subsystem of inflectional stems in the lexicon (sing vs. sang vs. sung; bring vs. brought; foot vs. feet; woman vs. women; etc.), and there are some attempts at agglutinative combinatorics (sing+s, sing+ing; bless+ed; house+s, etc.).

An interesting solution to this problem of mixed typology was later suggested by the gradient system developed by a leading figure in linguistic typology, Joseph Greenberg (1915–2001). Here is Greenberg's gradient model for morphological types in the world's languages (Greenberg, 1960):

1) M/W: (Morpheme per Word) = Synthesis index
2) A/J: (Agglutinative intra-word morph per Juncture) = Agglutination index

3) R/W: (Lexical Root per Word) = Compounding index
4) D/W: (Derivative affixes per Word) = Derivation index
5) I/W: (Inflectional morpheme per Word) = Inflection index
6) P/W: (Prefixes per Word) = Prefixation index
7) S/W: (Suffixes per Word) = Suffixation index, etc.

Table 2.1 is compiled from an interesting essay applying and revisiting the Greenbergian indexical model on the empirical data obtained from Maori, an Austronesian language difficult to classify because of its complex word patterns. The table shows once more how puzzling questions in linguistics are solved as soon as gradient patterns and quantified profiles are fathomed from data sets.

Table 2.1 Greenbergian morphological types in 13 languages, compiled from the data by Krupa (1966), Greenberg (1960), and Cowgill (2000), on the basis of 100-word samples.

	M/W	A/J	R/W	D/W	I/W	P/W	S/W
Maori	1.18	0.88	1	0.13	0.14	0.09	0.06
Vedic Sanskrit	2.56	0.08	1.1	0.49	0.97	0.19	0.24
Bengali	1.9	0.46	1.09	0.28	0.53	0.01	0.8
Old Persian	2.41	0.2	1.02	0.41	0.98	0.19	1.2
Modern Persian	1.52	0.34	1.03	0.1	0.39	0.01	0.49
Homeric Greek	2.07	0.1	1.01	0.21	0.85	0.06	1
Modern Greek	1.82	0.4	1.02	0.12	0.68	0.03	0.77
Old English	2.12	0.11	1	0.2	0.9	0.06	1.03
Modern English	1.68	0.3	1	0.15	0.53	0.04	0.64
Yakut	2.17	0.51	1.02	0.35	0.82	0	1.15
Swahili	2.55	0.67	1	0.07	0.8	1.16	0.41
Vietnamese	1.06	0	1.07	0	0	0	0
Eskimo	3.72	0.03	1	1.25	1.75	0	2.72

Eskimo and Vietnamese appear like antipodes, the former being 'hyper synthetic', weakly agglutinative, and rather strongly inflectional and derivational, the latter having minimal synthesis index (one morpheme = one word), and deprived of all the other properties. As extremes, these two languages can be held as antithetic paragons, or polar types. Maori clings more to the Vietnamese isolating type, although it is somewhat agglutinative, and Swahili shares more features with Eskimo than with other tokens of this sample. Its main idiosyncrasy lays in its prefixation index, which is much higher than any other language in this sample. Interestingly enough, ancient Indo-European languages, such as Vedic Sanskrit, Homeric Greek, and Old English have a more balanced regime of morphological properties, than for example, Yakut (an Altaic agglutinative language); they tend to be strongly synthetic, fairly agglutinative and inflectional, and somewhat derivational. However, in their modern counterparts (Bengali, Modern Greek, and Modern English), most properties have been eroded, except synthesis and inflection, which manage to gain ground to some extent, pointing at a mild inflectional drift.

When reading Table 2.1, the late nineteenth century Schleicherian evolutionary hypothesis, adapted from an initial intuition voiced by Franz Bopp, that languages would evolve from the isolating type to the inflectional type, through the agglutinative type, may now seem naïve [11]: or, at least, more than an understatement or a circular assessment, as many cycles of templatic expansion and compression or deduction of morphological templates may occur through space and time in linguistic stocks (see the interesting case of Sino-Tibetan languages from the Rgyalrongic and Kiranti genera, surveyed by Jacques [2016], which can nowadays be classified as polysynthetic. Questions can be raised about which type should be posited first: isolating or (poly)synthetic?) Greenberg's ratios of formative resources and strategies per word have gained stronghold in modern typology; the major database for linguistic typology available nowadays, the WALS (World Atlas of Language Structure, mentioned earlier), follow this logic, taking into account functional units and their distribution in words and phrases. Ratios for the number of phonemes per syllable, or affixes per word, canonical word order, the distribution of markers on the verb or the nouns or adjectives (head marking vs. dependent marking, a notion developed initially by Nichols [1986]) are among the main tenets of the method.

2.5.1 Epistemological Consequences: The Open Taxonomy Challenge

To what extent are language classifications reliable? What do we capture when working out a phylogramm or Stammbaum? Does it tell us anything about evolution in space and time of the linguistic family or the genus at stake? How can these results be correlated with external factors, and bring positive information about migrations, social networks; the whole bulk of questions involved in any evolutionary process, as mentioned earlier, can be slightly reformulated now under the following heads: geometric expansion, disparition of ancient clades, differentiation paths, inner structure of genealogical ramification, diverging vs. converging varieties, emergence of innovative varieties or clusters embedded in living or dead

---

[11] Leonard Bloomfield, in a paper that was among the most important ones of twentieth century linguistics, wrote in a note: *Bopp took for granted that the formative elements of Indo-European were once independent words; this is a needless and unwarranted assumption. The last descendant of his error is the assumption that Indo-European compound words are historically derived from phrases (Jacobi, Compositum und Nebensatz, Bonn 1897; this even in Brugmann, Grundrisz I, 1, pp. 37. 78; see also TAPA 45. 73 ff.). The notion is gaining ground that some forms have less meaning than others and are therefore more subject to phonetic change (Horn, Sprachkürper und Sprachfunktion, Palaestra 135, Berlin 1921); I, for one, can discover no workable definition of the terms "meaning" and "phonetic change" under which this notion can be upheld. The whole dispute, perhaps today as unstilled as fifty years ago, about the regularity of phonetic change, is at bottom a question of terminology* (Bloomfield, 1926, p. 153, note 3). As compared to this sharp criticism, our statement here about the naivety of the Schleicherian evolutive hypothesis of a path such as isolating > agglutinative > inflectional, can be viewed as a mild understatement. Nevertheless, one should be cautious in avoiding anachronism on this point: for Franz Bopp and August Schleicher, the evolutionary drift hypothesis on the three basic morphological 'types' was indeed heuristic; however, this hypothesis has often been connoted as ethnocentric, the inflectional Indo-European type being set above all others, as an accomplished product of evolution. Of course, this view is wrong, and clearly belongs to doxa and ideology, rather than to science.

clades, hierarchy of clades, the status of intermediate stages, extinction, survival and resilience of higher level clades.

For instance, the phylogram of Mayan languages in Figure 2.5 is based on a set of 70 cognates taken from the etymological dictionary by Kaufman and Justeson (2003), converted into a database (Léonard, 2014, p. 38).[12] The closer a Mayan language or variety stands from the root of the cladogram, the more 'conservative' the language is; the farther on the right, the more 'innovative'. Interestingly enough, a chronologically intermediate variety, available through the philological analysis of ancient Mayan script, namely EpM (Epigraphic Maya, a variety of Ch'olan), has been included, and stands as a key token before a complex cluster (indexed D in the cladrogram); it covers several distinct and geographically distant varieties, such as Western Huastec (or Tének) and some Ch'olan–Tseltalan and Q'anjobalan languages (CHJ & TOJ).

The procedure has elegantly captured the idiosyncrasy of the remainder of the Q'anjobalan genus (clade A, with MCH, TUZ on the one hand, and AKA, POP, QAN on the other), which cluster separately above. The position of the reconstructed (but nonpositively attested, in contrast to EpM) protolanguage such as pCh (Protocholan), next to its descendent CHL (Ch'ol) at the bottom of clade D introduces a slight mismatch, as epigraphic Mayan from the Classic is generally supposed to match more or less with proto-Ch'olan. However, this aporia can be considered heuristic as the relation between the two varieties.

The opposition and the inner structure of clades A and B raise a lot of issues for discussion: on the one hand, these clusters cover the bulk of Western Mayan languages, on the other hand, not only does it confirm the ambivalent status of Q'anjob'alan languages, at the crossroad between Western and Eastern Mayan[13] languages, but also the very intricate distribution of the languages which have been more in contact with the Ch'olan–Tseltalan cluster, such as TZE, TZO, CHJ, and TOJ. The distance to the root is also correct, with Soconusco Q'anjobalan languages such as MCH and TUZ and the Q'anjobalan nucleus AKA, POP, and QAN clustering slightly apart in clade A, while languages intensively involved in contact with Tzeltalan, as CHJ and TOJ are embedded in the middle range of clade D, close to Ch'olan and Tzeltalan languages. Clades E to J parse the Eastern Mayan languages (all of them spoken in Guatemala Highlands). The clear-cut subdivision between the two main genera, that is, Mamean vs. Quichean, is consistent with current classifications, and the inner structure of Mamean in clade E is flawless, with Ixil (IXL) as an outer member of the clade, TEK and AWA associated with a contact dialect of Mam (MamI), and the remainder of Mam proper on a sub-clade of its own. The last clade at the bottom covers the Quichean genus, with two sub-genera: F (Poqom and Q'eqchi', in the Eastern part, with USP embedded through contact with Q'eqchi') vs. G (K'iche proper, with Tz'utujil vs. Kaqchikel, associated with the peripheral contact varieties Sipakapense and Sacapultek: SAK and SIP in the sub-clade I). Interesting details

---

[12] The fine grain details of the dialects can be found in Kaufman and Justeson (2003), but is hardly relevant here, for example, MAMc:Cajola Mam, MAMo:Ostuncalco Mam, MAMl:San Idelfonso Mam, MAMt:Tacana Mam. The design of the Mayan etymological database from Kaufman and Justeson's dictionary and the cladisting processing performed with PAUP was initially carried out within the framework of a more generic research on cladistics applied to linguistic families and genera, by the Musée de l'Homme (Léonard, Vulin, and Darlu 2009).

[13] See Mateo Toledo (1999) for a comprehensive and detailed survey of Q'anjob'alan diasystemic variables.

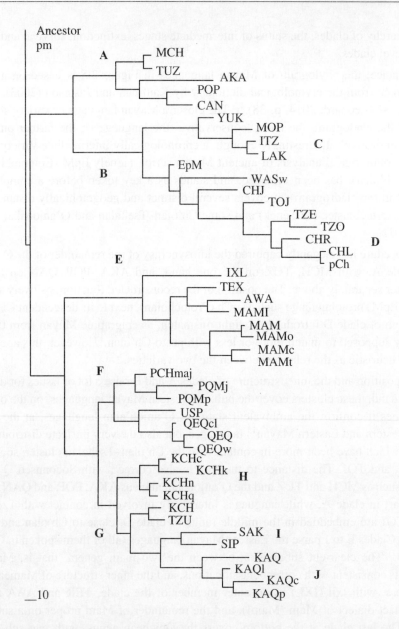

Figure 2.5 Phylogram of Mayan languages from the etymological dictionary by Kaufman and Justeson (2003) converted into a database (Léonard, 2014, p. 38). AKA: Akatek, AWA: Awakatek, CHL: Ch'ol, CHJ: Chuj, CHT: Chontal (from Tabasco), CHR: Ch'orti', ITZ: Itzaj, IXL: Ixil, KAQ: Kaqchikel, KCH: K'iche', LAK: Lakandon, MAM: Mam, MCH: Mocho, MOP: Mopan, PCH: Poqomchi', POP: Popti', PQM: Poqomam, QAN: Q'anjob'al, QEQ: q'eqchi', SAK: Sakapultek, SIP: Sipakapek, TEK: Teko, TOJ: Tojolab'al, TZE: Tzeltal, TUZ: Tuzantek, TZO: Tzotzil, TZU: Tz'utujiil, USP: Uspantek, YUK: Yukatek, WAS: wastek (Huastec, or Tének).

lie in the relationship between sub-clades I and J, as compared to H: why SAK and SIP cluster with Kaqchikel; why Tz'utujil lingers at the periphery of this conjunct, associated to K'iche'. Here too, many debates about contact trends within the Eastern branch of Mayan languages find, if not a definite solution, at least, heuristic detailed configurations, consistent with state of the art configurations in the domain of Mayan languages classification (see Law [2011] for an overview). Most of the challenging questions relevant to evolutionary linguistics are embodied in this phylogram. Its main purpose is precisely to bring new insights and to highlight taxonomic intricacy, rather than to freeze once and for all the overall configuration of such a complex language family like the Mayan, whose speakers, especially those who handled languages to be found in clade D, have dominated Southern Meso-America during the Classic period; whereas those included in clade C exerted some kind of hegemony, though under Aztecan rule, during the Post-Classic period.

All these external factors of distributed or shared hegemonies, or the emergence of scripts and koinés and even lingua francas, made the thread of town dialects, innovation centers, reaction–diffusion chains more complex. So did the contact, rise and fall of varieties, intercommunal alliances and rejection, pluralism, separation, segregation and assimilation trends over ages. All were dense and intricate phenomena, as it happened too in the Old World, so that complexity here cannot be understood only through quantitative or algorithmic approaches such as dialectometry or cladistics, even if the two methods are highly complementary, and give some positive results for historical interpretation.

As we just saw with the Q'anjobalan clade within the Mayan phylogenetic tree, some genera happen to be dialect continua,[14] so that their inner constituents may split and be 'attracted' by neighboring clades, in phylogenetic taxonomies. The inner structure of the communal aggregate (here, a diasystemic aggregate) may be scattered out of its nest, and gain taxonomic autonomy, as clade A, or cluster with contact languages in a strikingly discontinuous way as in the case of clade D, where clade C splits apart the Q'anjobalan continuum: this is especially impressive in the case of Chuj, while it could be expected from Tojolabal, for which it is hard to decide whether the language was initially Q'anjobalan, and became lately Tzeltalan, or the reverse (see Schumann [1983] on this particular issue).

Cladistics is all the more an interesting method to work out intricate taxonomies, as it is by no means a trivial quantitative method, summing up similarities and differences, like dialectometry, in a binary way (Goebl, 1981, 1984, 2002, 1998; Goebl and Schiltz, 1997). In Figure. 2.6, the reflexes of the etymon *jooloom, 'head' in Proto-Mayan (pM) are parsed in a graph, displaying the characters upon which pairs will be computed, with specific structural weight indexed on the paths (from 1 to 3 here) to measure cladistics distance, taking into account the structural 'cost' of each sound change. Computed characters (Gaillard-Corvaglia et al., 2007) bear indices in Greek letters. For instance, character (α) *jooloom*, pronounced [xo:lo:m], occurring in pM (Proto-Mayan) but also later, in Epigraphic Maya (EpM), matches the etymon, and amounts to a mere retention; whereas character (β) *joolon*, from *jooloom

---

[14] In a dialect continuum, a set of languages, spoken across a geographical area, are mutually intelligible locally (between neighboring varieties, which differ only slightly) but they may become quite different from each other as the mutual distance increases, so that widely separated varieties are not mutually intelligible.

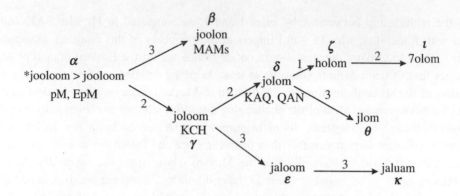

Figure 2.6 Graph of the reflexes of the etymon *jooloom ('head') in Proto-Mayan. From Léonard et al. (2010) (sfs.snv.jussieu.fr).

(*jooloom > joolon*), cumulates two processes: final vowel shortening (*jooloom > joolon*, value: 2) and final nasal sonorant delabialization (*jooloom > joolon*, value: 1). The path (γ) > (ε), *joloom > jalocm* shows a vowel shift (o > a), from a form which has already undergone initial vowel shortening, so that the previous weight will eventually cumulate to the heavy weight of this particular sound change (weight: 3), etc.

Clades A, C, E, F, G in the Mayan phylogram accounted for dialect continua; whereas, clade D included structurally very different languages. A similar discrepancy haunts language classifications, and make the task of linguistics all the more difficult. In the case of Mayan languages, except for a massive migration supposed to have taken place in ancient times, from the south to the north, which displaced Huastec (WAS), as suggested by its position in the D clade, populations were probably relatively stable before the Conquest and colonial times. However, what happens when communal aggregates have been migrating intensively through the ages? This is particularly the case of a genus within Altaic: Mongolian, which has undergone intense phases of military expansion through the Middle Ages (1187–1297), and thereafter, and pastoral nomadism throughout a vast region of Eurasia.[15] If the currently agreed classifications propose a set of isolates, considered as separate languages (Dagur, Moghol, and Monguor) in a threefold area, divided between a central area in the eastern side (Mongolian proper, including Khalkha), North Mongolian (Buriat), and West Mongolian (Oirat, including Kalmuck, according to some scholars), on the basis of multiple criteria, the Soviet school proposed only three major branches on the basic of a single criterion vowel harmony: the northern, or 'synharmonic' area, with Mongolian, Buriat, Kalmuck, and Oirat; the southeastern or nonsynharmonic area (Dagur, Monguor, Santa, and Baoan), and the 'intermediate' area, with Khalkha and Moghol.

---

[15] See Binnick (1987) for a comprehensive and detailed account of the attempts at classifying the Mongolian language. The author enumerates 28 structural variables, mostly from phonology and morphology, which give a complex, yet well-balanced overview of differentiation processes over the Mongolian diasystem. Poppe's (1955) work remains the basic reference to understand further the thread of Mongolian languages diversification in space and time.

This classification ostensibly agrees with external factors, history and, partly, the geography of the Mongolian expansion, but it relies on one, allegedly massive and powerful, isogloss: vowel harmony. Yet, complexity is lost, especially if we bear in mind the 28 phonological and morphological criteria suggested by Binnick (1987), or the comprehensiveness of Poppe's diachronic and diasystemic account of an even greater number of varieties and languages from this genus.

A cladistic processing of word lists available in Svantesson et al. (2005) for thirteen languages and dialects from the Mongolian genus, applied to the same number of cognates as for Mayan languages, is given in Figure 2.7. It converges with the previous fourfold division, with three separate languages in the outer clade: Kamnigan (Kmn). This small heterogeneous set (Moghol in India, as opposed to Dagur and Kamnigan in the northeastern periphery of the Mongolian area) stands as the peripheral class. The next clade, B, includes again a peripheral

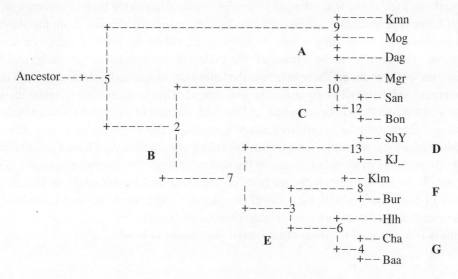

Figure 2.7 A cladistic phylogram for Mongolian languages. Realized by Pierre Darlu. Baa: Baarin, Bon: Bonan, Bur: Buriad, Cha: Chahar, Dag: Dagur, Hlh: Khalkha (Literary Mongolian), KJ: Kangjia, Kmn: Kamnigan, Mog: Moghol, Mgr: Monguor, San: Santa, ShY: (Butha) Shira Yugur.

one, yet much more homogeneous. In the C clade, southwestern dialects cluster together as small isolates in northern China (Monguor, Santa, and Bonan). Clade D contains two other languages spoken in roughly the same zone as languages in the previous class: Shira Yugur and Kangjia, which therefore are parsed discontinuously here in an additional peripheral southwestern clade. Clade F also contains discontinuous languages: Kalmuck, located far in the West, on the shores of the Caspian sea, in the Lower Volga basin, clustering with Buriat, a dialect spoken north of Mongolia. Once more, no clearcut geographic pattern—nor ethnic heritage, as some of these languages belong to the Oirat group, like Kalmuck, and others do not—can be found in this distribution. Clade G does not even really match the notion of Central Mongolia, as it gathers Khalkha, along with Chahar and Baarin, making up a somewhat more consistently southeastern peripheral group. Nevertheless, the heterogeneity of the phylogram still teaches us a lot about the inner structure of the Mongolian genus: instead of an organic, dense or compact diasystem, it is a centrifuge matrix, partly made of condensated typological clusters, as clades C and D in the southeast, and clade G in the central and southeastern area, partly made of heterogeneous classes of far reaching spreading types, such as A and F. The intense mobility of some expansive communal aggregates in the outskirts of a vast central zone occupied by languages of the G clade in the south/southeastern part alternate with mobile segments as a consequence of long-range migrations and resettlements. These language dynamics can be considered as more relevant and heuristic than taxonomies relying upon too massive typological criteria, such as synharmonic vs. nonsynharmonic vowel systems, or the mingling of clanic (such as being or not being Oirat) and geographic criteria.

The phylogram in Figure 2.7 enhances the difference between 'attributive dialect' vs. 'blind dialect' (Nerbonne and Kretzschmar, 2003): the former is sustained by individual or collective representations and consent, attributing properties or affiliation on the basis of common sense or current knowledge available; while the latter, though potentially discrete from the standpoint of its structure and form, may remain inconspicuous, below the level of public or scientific awareness. The former can be termed as 'the dialect Smith speaks' or 'the dialect of South Boston', or 'an Oirat dialect' here; whereas the latter may be ignored as long as we can ignore, as Nerbonne and Kretzschmar put it, 'the question of what is distinctive to some groups of varieties'. Blind dialect can be retrieved, and the intricate web of attributed factors which make it hard to fathom, may be unraveled through complexity theory procedures, as a direct or a byproduct of self-organization. This alternative standpoint, in turn, may have a positive impact on the advancement of knowledge, and on the understanding of contextual ontologies in space and time. If we are to integrate the methods of linguistics and the sociology of language into a theory of higher scope, with the tools of complexity theory, we need open taxonomies and much more flexible frameworks, knowledge objects, and results.

This is indeed one of the main objectives of the volume at hand.

Chapter 3

Comparison Based on String Metric

A *string metric* is any metric distance between entities which can be associated with a string. String metric-based methods have been developed and used for tackling various problems, from plagiarism detection and DNA/RNA analysis, image analysis and recognition, to data mining and integration, and incremental search, to name a few. In this chapter, we consider some simple examples of metric distances and apply them to some real examples related to language.

3.1 Levenshtein Distance

The most widely known string metric for measuring the difference between two sequences is the *Levenshtein distance*, also known as *edit distance*, named after Vladimir Levenshtein, who considered this distance in 1965 (Levenshtein, 1966).

The Levenshtein distance $L(a, b)$ between two given strings a and b, each composed of a set of characters, is defined as the *minimum* number of edit operations, including character addition, removal, and replacement, needed to turn a into b or vice versa (see e.g. Apostolico and Galil [1997]). Here is an example of how the Levenshtein distance can be used.

Example: Levenshtein distances between three given words. Let us consider three different locations in the Basque countries, labeled here with $k = 1, 2, 3$, where three correspondingly different dialects of Basque are spoken, and compare the three variants of the same word, the Basque word for 'I am', in these locations. The words are $a_1 = $ *naiz*, $a_2 = $ *nais*, and $a_3 = $ *nas*. Comparing these three words with each other, we notice the following relations:

- $a_1 = $ *naiz* vs. $a_2 = $ *nais*: *naiz* $\rightarrow$ *nais* by one replacement operation $z \rightarrow s$; thus, $L_{12} = L(naiz, nais) = 1$.
- $a_1 = $ *naiz* vs. $a_3 = $ *nas*: *naiz* $\rightarrow$ *nais* $\rightarrow$ *nas* by two edit operations: replacement $z \rightarrow s$ and deletion of *i*; so $L_{13} = L(naiz, nas) = 2$.
- $a_2 = $ *nais* vs. $a_3 = $ *nas*: *nais* $\rightarrow$ *nas* by deletion of *i*; $L_{23} = L(nais, nas) = 1$.

The Levenshtein distances thus obtained can be organized into a *distance matrix* $L = \{L_{ij}\}$, in this example, with 3 words, a 3×3 matrix ($i, j = 1, 2, 3$). The diagonal elements are equal to zero ($L_{ii} = 0$, since any word is identical to itself) and symmetrical (because a metric distance is symmetrical, $L_{ij} = L_{ji}$),

$$\{L_{ij}\} = \begin{pmatrix} L_{11} & L_{12} & L_{13} \\ L_{21} & L_{22} & L_{23} \\ L_{31} & L_{32} & L_{33} \end{pmatrix} = \begin{pmatrix} 0 & 1 & 2 \\ 1 & 0 & 1 \\ 2 & 1 & 0 \end{pmatrix}.$$

The computation of the Levenshtein distance is simple for short strings with small differences; but in more complicated cases, the optimal sequence of changes associated with the minimum number of edit operations may not be obvious. In the analysis of large databases, it is convenient, or necessary, to use a numerical code for computing the Levenshtein distances (Apostolico and Galil, 1997).

The popularity of the Levenshtein distance is justified by the simplicity of its definition, the clear meaning, and the fact that it is straightforward to use. However, these same features represent obvious limits when addressing more precise questions, since the Levenshtein distance is independent of the following factors:

- The type of operation
- Which characters are involved (e.g., whether $a \to e$ or $a \to i$)
- The order of operations
- Possible correlations between edit operations (e.g., a transposition of two syllables would be detected as a series of independent edit operations on single characters),

All these factors are relevant aspects from a linguistic perspective. Some general properties of the Levenshtein distance are the following.

(i) It is at least the difference of the sizes of the two strings.
(ii) It is at most the length of the longer string.
(iii) It has the *Hamming distance* (explained later in this chapter) as an upper bound if the two strings compared have the same size.
(iv) It is positive or zero (being zero if and only if the two strings are equal).
(v) It is symmetrical (it does not depend on the order of the two strings).
(vi) It satisfies the *triangular inequality*: the Levenshtein distance between two string a and b cannot be greater than the sum of the Levenshtein distances of the two strings from a third string c, that is, $L_{ab} \leq L_{ac} + L_{bc}$.

The last three properties define the Levenshtein distance as a *metric*, a relevant condition for setting up a consistent and reasonable distance-based criterion to evaluate mutual distances. However, the Levenshtein distance is not the only quantity which can be used for measuring the similarity of strings (see Apostolico and Galil [1997]).

3.2 Normalized Levenshtein Distance

The basic Levenshtein distance defined in the previous section may have a certain bias since longer words have a higher probability of undergoing changes than the shorter ones. Thus, the Levenshtein distance has an explicit dependence on the length of the word analyzed, a bias that can be corrected, for example through the *relative Levenshtein distance* L_{rel}, obtained by dividing the standard Levenshtein distance by the length of the longer word. This is the Levenshtein distance which will be used in the following sections,

$$L_{\text{rel}}(a_1, a_2) = \frac{L(a_1, a_2)}{\max(|a_1|, |a_2|)}, \qquad (3.1)$$

where a_i ($i = 1, 2$) are strings and $|a_i|$ represents the length of a_i (the number of characters). The normalized distance is a number in the interval $L_{\text{rel}} \in (0, 1)$. Notice that there are different ways to normalize the Levenshtein distance (see e.g. Wichmann et al. [2010]).

3.3 Levenshtein Distance for Language Comparison

The example of the Levenshtein matrix for $N = 3$ words with the same meaning, discussed in Section 3.1, can be generalized for comparing languages with each other. In order to estimate the distance between two given languages i and j, here we consider the more general example of N dialects or languages spoken in N different corresponding locations, in which a set of M different words are recorded, for example, corresponding to M assigned semantic meanings in a lexicological analysis (we assume that all M words are recorded in all the N locations). The final database can be compiled in the form of a matrix $A = \{a_i^{(k)}\}$, consisting of strings with a subscript $i = 1, \ldots, N$, running over the different locations/languages, and a superscript $k = 1, \ldots, M$, labeling the M different semantic meanings; thus, the generic string $a_i^{(k)}$ represents the word with the kth meaning recorded in the ith location.

A simple estimate of a mean (metric) distance between language i and language j is the arithmetic average $\bar{L}_{ij}$ of the Levenshtein distances between all the pairs of words—one word in location i and the other word in location j—with common semantic meaning,

$$\bar{L}_{ij} = \frac{1}{M} \sum_{k=1}^{M} L_{\text{rel}}(a_i^{(k)}, a_j^{(k)}). \qquad (i, j = 1, \ldots, N) \qquad (3.2)$$

This formula should be corrected in the case where for a given pair of locations i and j, not all the M pairs of words with the same meaning are available but only a smaller number $M_{ij} < M$, by simply replacing M with M_{ij} in Eq. (3.2) and of course limiting the sum to the M_{ij} pairs available. In turn, the set of average Levenshtein distances $\bar{L}_{ij}$ thus computed for each pair of locations (i, j) can be organized into another $N \times N$ matrix, the Levenshtein matrix $L = \{\bar{L}_{ij}\}$. The Levenshtein matrix L lends itself to informative and complementary representations and visualizations that can provide precious information about the structure of the language family under study, as illustrated by the examples in the following sections.

Another generalization is to maintain the basic definition of the Levenshtein distance but to weight the data considered. For example, when computing an average Levenshtein distance $\bar{L}_{ij}$ as in Eq. (3.2), the single data types k (not the basic edit operations) have to be assigned different statistical weights p_k. Then the average Levenshtein distance becomes

$$\bar{L}_{ij}^{(w)} = \frac{\sum_{k=1}^{M} p_k L_{\text{rel}}(a_i^{(k)}, a_j^{(k)})}{\sum_{k'=1}^{M} p_{k'}}. \tag{3.3}$$

The average has to be explicitly normalized by the sum of the weights unless they represent probabilities normalized to one, that is, $\sum_k p_k = 1$. The arithmetic average in Eq. (3.2) corresponds to the statistical weights p_k all being equal to each other.

The set of possible language changes is wide and in many cases, the Levenshtein distance represents a basic approximation. For this reason, other measures of edit distance have been introduced. For instance, the basic edit operations considered in computing a distance can be treated in different ways, contributing in different ways to the final value of the distance.

A generalized distance can be defined considering different sets of edit operations that suit particular problems: we mention the Damerau–Levenshtein distance, which also includes *transposition* of two adjacent characters (besides insertion, deletion, and substitution) as an elementary edit operation: if the basic Levenshtein distance is used, one would need two edit operations to make one transposition.

On the other hand, other measures of distance with a more restricted set of operations could also be useful in some problems. A most relevant example is the *Hamming distance*, which allows only substitutions and hence is suited for comparing strings of the same length (the Levenshtein distance between strings of equal length reduces to the Hamming distance). The Hamming distance—or distances basically equivalent to it—appears in many different types of analyses, although it is not referred to with its name. For example, one can define a language in terms of a set of selected linguist fetaures. This can be done by assigning a bit s_k to each feature and defining each language through the bit string $s = (s_1, \ldots, s_F)$, where the F bits, assuming either the value 0 or 1, represent, for example, the presence/absence of F linguistic feature. In a more general case, there can be more than two variants, and the bits b are assigned to variables assuming Q possible values, so that there are F^Q possible languages. In this case, the Levenshtein distance or the Hamming distance provides a measure of the distance between two strings (languages).

Other examples of distance are the *longest common subsequence* distance, which allows insertions and deletions but not substitutions, and the *Jaro distance*, which allows only transpositions.

3.4 Similarity Index

Once one has defined a normalized distance, it is natural to define a corresponding normalized similarity index as follows.

$$s(a_1, a_2) = 1 - L_{\text{rel}}(a_1, a_2). \tag{3.4}$$

This equation can serve equally well for a description of the relations between languages. Like the normalized Levenshtein distance, the similarity index is a number in the interval $s \in (0, 1)$.

For any other types of distances normalized in the interval (0, 1), one can define a similarity index s,

$$s = 1 - d. \qquad (3.5)$$

The normalized distance d and the similarity index s can be considered as equivalent and complementary quantities.

In order to investigate a language family, we need only a relative scale of similarities and differences, not their absolute values. Thus, the procedures outlined earlier are quite general in nature and can be used whenever a set of distances has to be estimated according to some (dis)similarity criterion, while the choice of the metric-based method is rather free.

However, relevant differences do turn up in some cases. When one is interested in finer details or different viewpoints to look at a system, it can be very informative to compare the outcome of analyses carried out on the basis of different criteria (see the case study of the Mazatec language in Section 3.5 and Tseltal in Section 3.7). In other cases, when the database is not so large, it can be noticed that suitable weighted averages using a restricted sample can provide richer information than arithmetic averages of larger samples (see the case study of the Basque language in Section 7.7).

3.5 Case Study: The Mazatec Language

In this section, we introduce the case study of the Mazatec dialects (used also in later sections as a working example), to illustrate how the concepts of edit distance and networks introduced in the previous sections can be further elaborated into methods and criteria suited for extracting information about a set of languages and their mutual relationships (Léonard et al., 2017).

Figure 3.1 The Mazatec dialects considered in our study. The 12 Mazatec dialect locations considered here coincide with those studied by Kirk (1966). (Map data © 2019 Google, INEGI.)

Table 3.1 Mazatec locations and abbreviations. Parts of names sometimes omitted are in square brackets [·]; alternative names are in parentheses (·).

1.	AY	[San Bartolomé] Ayautla
2.	CQ	[San Juan] Chiquihuitlán
3.	DO	Santo Domingo [del Río]
4.	HU	Huautla (de Jiménez)
5.	IX	[San Pedro] Ixcatlán
6.	JA	Jalapa [de Díaz] (San Felipe)
7.	JI / AS	[Santa María de la] Asuncion ([Santa María de] Jiotes)
8.	LO	San Lorenzo [Cuaunecuiltitla]
9.	MG	[San Miguel] Huautla
10.	MZ	Mazatlán [Villa] de Flores (San Cristobal)
11.	SO	[San Miguel] Soyaltepec
12.	TE	[San Jerónimo] Tecoatl

3.5.1 Mazatec: Introduction

Mazatec is an Otomangue language that is spoken in the southeast region of Mexico (SDSH, 2011–2016; Gudschinsky, 1955, 1958). Mazatec dialects are spoken by a population larger than 200 000 people, with a highly heterogeneous culture and a geographically diversified landscape. Mazatec dialects are a well studied subject of dialectology; this is because in microscale, they can provide a highly complex panorama comparable with, for example, the one of European languages (SDSH, 2011–2016; Gudschinsky, 1955, 1958, 1959; Kirk, 1966; Jamieson et al., 1988; Capen, 1996; Léonard et al., 2012; Kihm, 2014).

In this section, we will discuss the Mazatec dialects by analyzing some previously created databases related to the average linguistic Levenshtein distance between dialects (Heeringa and Gooskens, 2003; Bolognesi and Heeringa, 2002; Beijering et al., 2008) and others based on the mutual intelligibility of dialects (Kirk, 1970). Furthermore, we consider geographical and ecological constraints, as well as the social and economic conditions. The dialects will be visualized in the form of various networks corresponding to different aspects of the Mazatec dialects. This section is based on Léonard et al. (2017, 2019), where one can find a more detailed discussion on the Mazatec dialects.

The Mazatec language will be discussed again in Chapter 6 in connection with a simple mathematical model of the Mazatec expansion.

3.5.2 Mazatec: A short overview

Mazatec is an endangered language that has survived due to a relevant popoulation size (more than 200,000 speakers) and emerging language engineering for literature and education through modern spelling conventions. However, it is still very vulnerable, as suggested by the ALMaz.[1]

---

[1] A Linguistic Atlas of Mazatec, see Léonard et al. (2012).

The Mazatec area, located in the middle of the Papaloapam Basin, enjoys a smooth transition between the plains in the east and the mountains in the west (see Figure 3.1). This ecologically strategic position turned out to be fatal to the Mazatec Lowland, which was partly drowned by the Miguel Alemán Dam in the 1950s. The drowning of the Mazatec Lowland caused the collapse of the agrarian system based on coffee crops and cooperatives in one of the few regions where native peasants worked their own *microfundio*, and disrupted patterns of cross-regional integration evolved since the Olmec times (Killion and Urcid, 2001). Furthermore, a constant migration flux toward urban centers has affected the Mazatec community. All these factors have contributed to the vulnerability of the Mazatec language.

Table 3.1 lists abbreviations and names of the geographical locations (and corresponding dialects) considered in this study. For convenience, the Mazatec diasystem (Popolocan, Eastern Otomanguean) can be divided further into the different regions listed in Table 3.2 (see also Léonard and Fulcrand [2016]).

Table 3.2 Regions of the Mazatec dialects.

Highland complex	*Central Highlands*	HU—Huautla de Jiménez JI/AS—Santa Maria de la Asuncion MG—San Miguel Huautla
	Northwestern Highlands	TE—San Jeronimo Tecoatl LO—San Lorenzo
Lowland complex	*Eastern Lowlands*	SO—San Miguel Soyaltepec
	Central Lowlands	IX—San Pedro Ixcatlán
	Piedmont or Midlands	AY—Ayautla JA—Jalapa de Diaz DO—Santo Domingo
Periphery		MZ—Mazatlán Villa de Flores CQ—San Juan Chiquihuitlán

Areas and subareas can also be defined in terms of percentage of Mazatec speakers, as in Figure 3.2. The Mazatec language is still used intensively in the core of the Mazatec area, where spots densely cluster (see Figure 3.2).

3.5.3 Levenshtein analysis of Mazatec dialects

In this section, we will compare Mazatec dialects with each other by computing pair-wise the average Levenshtein distances between dialects. The results presented here are obtained by analyzing phonological and morphological patterns, while etyma are not used, contrary to a phylogenetic approach. These results are expected to highlight the ontological distances and be able to characterize the complexity of the links between dialects.

We proceed as described in Section 3.1 (in particular using Eq. [3.2]), using the dialectological data obtained by Kirk (1966) (see Léonard et al. [2017] for more details). The values of the average Levenshtein distance are collected in Table 3.3. The network is visualized

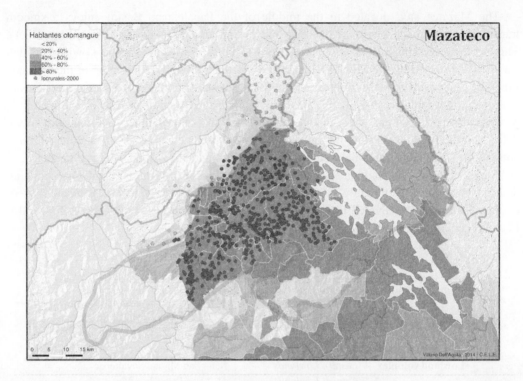

Figure 3.2 Vitality scale of Mazatec areas measured by the percentage of Mazatec speakers (see legend). Spots represent urban centers. (CELE [Vittorio dell'Aquila, 2014]. Official census data [2002].)

in Figures 3.3–3.7 for various values of the threshold T on the Levenshtein distance; it is plotted containing only links between dialect i and dialect j which satisfy the inequality

$$\bar{L}_{ij} < T. \tag{3.6}$$

For $T = 0$, there are no nodes that are linked to each other; increasing the value of T, more and more links between dialects nodes appear, creating a network. For the value $T = 1$, all the dialects are connected to each other, producing a fully connected network. However, at any value of T, one can effectively visualize the relative strengths of the links between dialects by drawing each link with a smaller (bigger) thickness and lighter (darker) color, the larger (smaller) the corresponding Levenshtein distance is.

For $T = 0.20$, a choreme appears, which can be considered as the kernel area (see Goebl, 1998, p. 555). The link joining JA and DO is plotted as a bolder line, meaning a smaller Levenshtein distance, which suggests the existence of a separate dialect.

For $T = 0.22$, the [HU–JI] choreme becomes visible in the Highlands. Furthermore, the group [AY[IX[DO–JA]]], now including AY, appears.

Table 3.3 Levenshtein distances between pairs of Mazatec dialects for the 12 locations studied by Kirk (1966). 111 cognates, including a sample of all lexical categories. Data processing: CELE, V. dell'Aquila (2014).

	AY	CQ	DO	HU	IX	JA	JI	LO	MG	MZ	SO	TE
AY	0	0.28	0.20	0.32	0.21	0.24	0.30	0.52	0.29	0.27	0.24	0.29
CQ	0.28	0	0.30	0.38	0.30	0.33	0.37	0.54	0.34	0.35	0.30	0.34
DO	0.20	0.30	0	0.33	0.19	0.11	0.33	0.54	0.27	0.26	0.24	0.28
HU	0.32	0.38	0.33	0	0.32	0.30	0.21	0.53	0.25	0.30	0.24	0.33
IX	0.21	0.30	0.19	0.32	0	0.22	0.31	0.53	0.29	0.27	0.24	0.25
JA	0.24	0.33	0.11	0.30	0.22	0	0.32	0.55	0.28	0.28	0.25	0.28
JI	0.30	0.37	0.33	0.21	0.31	0.32	0	0.55	0.33	0.28	0.24	0.28
LO	0.52	0.54	0.54	0.53	0.53	0.55	0.55	0	0.55	0.33	0.50	0.50
MG	0.29	0.34	0.27	0.25	0.29	0.28	0.33	0.55	0	0.25	0.24	0.31
MZ	0.27	0.35	0.26	0.30	0.27	0.28	0.28	0.33	0.25	0	0.22	0.29
SO	0.24	0.30	0.24	0.24	0.24	0.25	0.24	0.50	0.24	0.22	0	0.26
TE	0.29	0.34	0.28	0.33	0.25	0.28	0.28	0.50	0.31	0.29	0.26	0

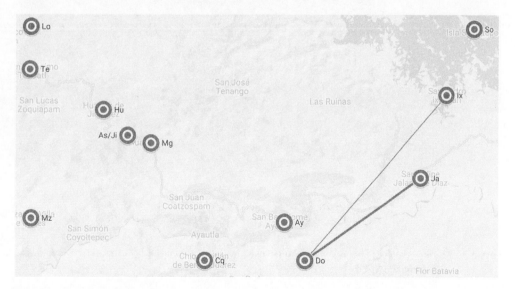

Figure 3.3 Mazatec dialect networks with normalized Levenshtein distances $L < T = 0.20$. (Map data © 2019 Google, INEGI.)

At $T = 0.24$, the group [MZ–SO] and the larger group [AY[IX[DO–JA]]] appear, in agreement with the 1955 Gudschinsky's model initially elaborated out of lexicostatistics (see Gudschinsky [1955, 1958, 1959] for details).

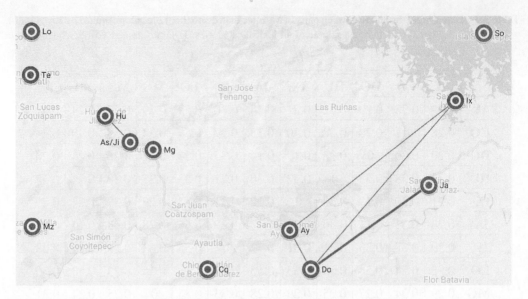

Figure 3.4 Mazatec dialect networks with normalized Levenshtein distances $L < T = 0.22$. (Map data © 2019 Google, INEGI.)

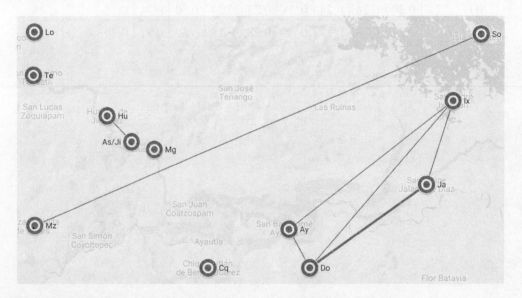

Figure 3.5 Mazatec dialect networks with normalized Levenshtein distances $L < T = 0.24$. (Map data © 2019 Google, INEGI.)

At the threshold value $T = 0.27$, various new groups appear, such as [TE[SO[IX]], [HU–JI–MG[SO]], and a macro-chain connecting MZ with the [IX–DO–JA–AY] choreme. At this threshold, only LO and CQ, the peripheral Mazatec varieties spoken in the northwestern Highlands and in the southwestern periphery, respectively, are still disconnected.

Comparison Based on String Metric

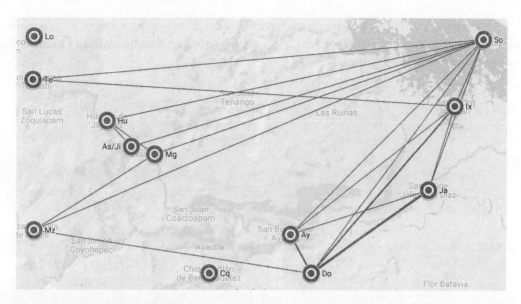

Figure 3.6 Mazatec dialect networks with normalized Levenshtein distances $L < T = 0.27$. (Map data © 2019 Google, INEGI.)

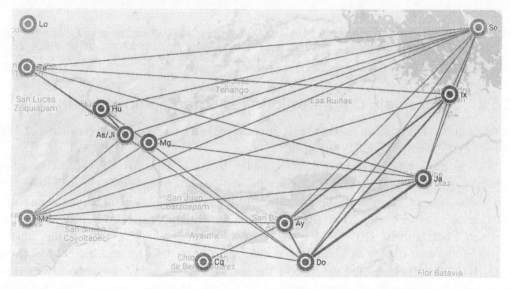

Figure 3.7 Mazatec dialect networks with normalized Levenshtein distances $L < T = 0.29$. (Map data © 2019 Google, INEGI.)

The closest correlation of CQ is eventually revealed by the network with threshold $T = 0.29$, showing a [CQ–AY] correlation. The reason why CQ connects to AY is the transitional status of AY, between the Lowlands and the Highlands, instead of being a structural heritage.

All the nodes, including LO, are connected using a neighbor graph, in which, by definition, each node has at least one link (for each node, only the strongest link, the shortest Levenshtein distance, is drawn) (see Figure 3.8).

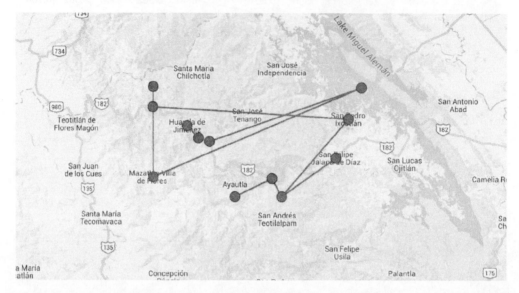

Figure 3.8 Nearest neighbor network based on the LD distances of 117 cognates, based on the data given by Kirk (1966). (Map data © 2019 Google, INEGI.)

3.5.4 Levenshtein distances for a restricted sample: Nouns

In this section, the results of another analysis are presented using a database consisting of 311 nouns (see Léonard et al. [2019] for details).

The corresponding Levenshtein distance matrix, shown in Table 3.4, was used to construct networks with different Levenshtein distance thresholds, shown in Figure 3.9. Such networks are similar but not identical to those obtained in the previous section. The first significant changes occur at threshold $T = 0.45$, when two choremes appear: [JA–IX] located in the southern Lowlands and [HU–JI–MG] in the central Highlands. The latter group creates a connection with dialect MZ, already considered to be ambivalent by Gudschinsky, in a [MZ[HU–JI–MG]] chain. On the base of the previous analysis, MZ would be expected to cluster at a later stage of structural identification with the Highlands choreme. This behavior suggests consistent word loans between MZ and the central area of the Highlands. The new network, appearing at the threshold value $T = 0.59$, crosses the whole area from west, where it is rooted in MZ, to the east, linking SO, with two lateral extensions toward the northwest (TE) and southeast (IX). The other network corresponding to $T = 0.72$ extends as far as AY and CQ.

Comparison Based on String Metric

Table 3.4 Levenshtein distance matrix based on nouns from Kirk's database: 311 nouns overall (Kirk, 1966).

	AY	CQ	DO	HU	IX	JA	JI	LO	MG	MZ	SO	TE
AY	0	0.632	0.629	0.668	0.606	0.607	0.636	0.981	0.562	0.573	0.582	0.708
CQ	0.632	0	0.717	0.703	0.666	0.704	0.589	0.978	0.627	0.645	0.636	0.688
DO	0.629	0.717	0	0.689	0.585	0.334	0.643	1,000	0.608	0.639	0.620	0.703
HU	0.668	0.703	0.689	0	0.593	0.655	0.346	0.897	0.402	0.481	0.519	0.550
IX	0.606	0.666	0.585	0.593	0	0.599	0.616	0.937	0.574	0.639	0.519	0.586
JA	0.607	0.704	0.334	0.655	0.599	0	0.617	0.945	0.594	0.604	0.585	0.675
JI	0.636	0.589	0.643	0.346	0.616	0.617	0	0.841	0.377	0.426	0.462	0.502
LO	0.981	0.978	1,000	0.897	0.937	0.945	0.841	0	0.883	0.892	0.884	0.870
MG	0.562	0.627	0.608	0.402	0.574	0.594	0.377	0.883	0	0.446	0.490	0.539
MZ	0.573	0.645	0.639	0.481	0.639	0.604	0.426	0.892	0.446	0	0.511	0.567
SO	0.582	0.636	0.620	0.519	0.519	0.585	0.462	0.884	0.490	0.511	0	0.574
TE	0.708	0.688	0.703	0.550	0.586	0.675	0.502	0.870	0.539	0.567	0.574	0

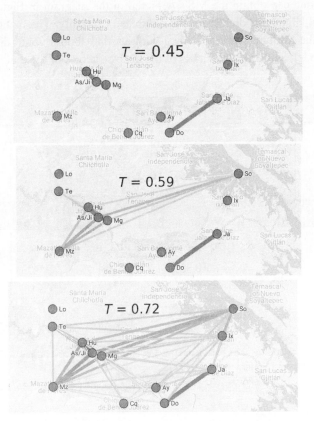

Figure 3.9 Dialect networks resulting from the average Levenshtein distance between nouns from Kirk's database, for different thresholds T. (Map data © 2019 Google, INEGI.)

3.5.5 Minimum spanning tree

Given a weighted network with N nodes, the *minimum spanning tree* (MST) is a particular spanning tree (see Section 2.5) of that network, obtained as the spanning tree with the minimum sum of all the edge weights (minimum total weight). In the present case, there are $N = 12$ nodes, corresponding to the 12 Mazatec locations. Therefore, the MST has 11 links.

In general, when using a threshold T for the Levenshtein distance to represent the corresponding language group, one may obtain a network that is not connected. In particular, one can obtain isolated nodes, as in the networks considered here, for which the representation offers no new information. A useful point in visualizing a network through the MST is that all the nodes of the networks are connected in a single network and one can have a general idea of the relative strengths of their links. The construction of an MST can be based on any type of distance known between nodes.[2]

Using the noun-based Levenshtein distance matrix of the Mazatec dialects discussed in the previous section, one can obtain the MST depicted in the diagram in Figure 3.10. The MST endows the central Highlands dialect JI with enhanced centrality. The fact that the transitional variety of AY in the Midlands is intertwined with another 'buffer zone' dialect, according to Gudschinsky's model, confirms the details of the deep structure of the dialect network.

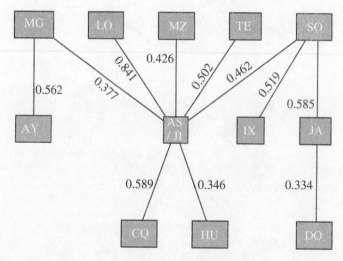

Figure 3.10 Minimum spanning tree based on the Levenshtein distance applied to nouns in Kirk's data. (Adapted from Léonard et al. [2017].)

---

[2] Among the standard ways to compute an MST, we mention (1) the *Kruskal algorithm*, which extends the minimum spanning tree by one edge at every discrete time interval by finding an edge which links two separate trees in a spreading forest of growing minimum spanning trees; (2) the *Prim algorithm* (used here), which extends the minimum spanning tree by one edge at every discrete time interval by adding a minimal edge that links a node in the growing minimum spanning tree with one other remaining node. The MST presented here was obtained using the inbuilt MST function in MATLAB (see Matlab documentation for details).

3.5.6 Dendrogram

A dendrogram is a diagram that represents a tree network (see Section 2.5). It is used, for example, in philogenetics, as a philogenetic tree, and in general to visualize the hierarchical clustering of a set of given nodes based on their mutual distances (many different types of distance can be used to this end). In the present case, we consider the dendrogram in Figure 3.11, obtained for the 12 Mazatec dialects considered as the nodes—starting from their mutual average Levenshtein distances in the matrix in Table 3.4.[3]

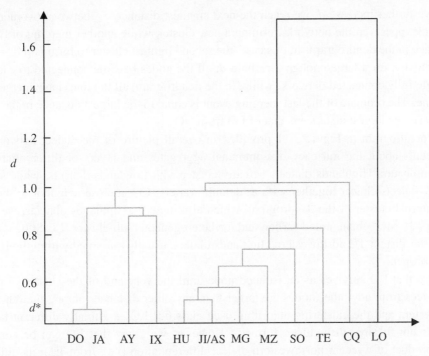

Figure 3.11 Levenshtein distance applied to nouns in Kirk's data: Dendrogram. (Adapted from Léonard et al. [2017].)

In a dendrogram, the vertical axis represents the distance d between clusters, while the horizontal axis only labels nodes and clusters; notice that for the sake of visualization, the order of labels on the horizontal axis is usually chosen based on the order in which nodes merge for increasing d. The dendrogram in Figure 3.11 compares the network of the Levenshtein distance matrix with varying threshold T (here referred to as d) used in the previous sections.

The nodes with the minimum Levenshtein distance d^* are JA and DO. Thus, using a threshold $d < d^*$, the network visualized contains only nodes disconnected from each other; correspondingly, in the dendrogram in Figure 3.11, a point is plotted at the abscissa of each dialect with a given ordinate d.

---

[3] Distances are not normalized here. The inbuilt function in MATLAB (see MATLAB documentation) was used to generate the hierarchical binary cluster tree (dendrogram).

When the threshold d is increased, nothing happens as long as $d < d^*$: the network does not change and, in the dendrogram, new points are plotted at the new d value corresponding to the same 12 abscissae of the dialects; in this way, the vertical lines representing the dialects in Figure 3.11 begin to form.

At $d = d^*$, the link connecting DO and JA appears in the network, creating the first cluster [DO–JA], and, in the dendrogram, the two lines corresponding to JA and DO are merged into a single line (which afterwards, for convenience, has an abscissa intermediate between the two original lines), representing the [JA–DO] cluster.

As we further increase d, we reach the next smallest distance d^{**} between two nodes, and a new link appears in the network, forming a new cluster, while another merging of two lines takes place in the dendrogram. In this case, the second tightest cluster to form is [HU–JI].

Eventually, for a large enough threshold d, all the nodes become connected to each other in a single fully connected network; while, in the dendrogram, all the lines have merged into a single line. The ordinate of the last merging event is equal to the largest distance in the system. In this case, the largest distance is that of LO from DO.

The dendrogram in Figure 3.11 provides an overall picture of the dialect network, with more details about the intricacy of communal aggregates and layers of differentiation. In this Stammbaum, Highlands dialects actually cluster with Lowlands dialects, while southern Midland dialects cluster together with a 'default' variety, CQ, a close neighbor in the south. In the internal cluster of the dendrogram, which also includes Highlands dialects, we see the [MZ–HU–JI–MG] chain, noted earlier, and the heterogeneous sub-cluster [IX–SO], associated with the far distant TE northwestern Highlands dialect, usually classified within the Highland dialects proper.

Notice that LO survives as an isolated node until the very end of the clustering process. Its distances from any other nodes are larger than any other distances among the other nodes in the system and the clustering algorithm used classifies LO as a totally different language that does not fit any other cluster formed. However, we know that LO can be considered as a byproduct of a recent northwestern dialect differentiation (i.e., from TE); its differences are indeed phonologically conspicuous because of a recent vowel shift, I → e, e → a, a → o, u → ï. The large distances of LO are probably due to the inability of the Levenshtein distance to distinguish between such a large vowel shift and an actual, deeper difference. Apart from the mentioned shortcoming, the dendrogram confirms the Mazatec tripartition [Midlands[Highlands–Lowlands]] and the [MZ[HU–JI–MG]] chain.

3.5.7 Multidimensional scaling

Multidimensional scaling (MDS) is another method that employs a defined distance for analyzing and representing the internal similarities and differences of large databases (Borg and Groenen, 2005). It can be applied to any type of distance, for example, the Levenshtein distance between words, the metric distance in a geometrical picture, or the geographical distance in a map.

In this method, initially, each element (here each dialect) is associated with a point in a multidimensional space, with a number of dimensions N, in general, comparable with the number of elements of the database. MDS uses a subspace with a small(er) number of dimensions D (usually with $D \ll N$) to provide an approximate but more simple and intuitive picture of the data.

Performing a comparison between dialects, MDS is used to locate the different dialects in a space $\mathcal{R}^D$ based on the strength of the links between them. Often, the embedding dimension is chosen as $D = 2$, for the sake of effective visualization. In the reduced space $\mathcal{R}^D$, similar dialects are close to each other while different dialects are far apart (Borg and Groenen, 2005).

In order to construct the MDS representation, one starts from the Levenshtein distance matrix $\{L_{ij}\}$. For a given D, the goal of the MDS method is to generate N D-dimensional vectors $x_1, ..., x_N \in \mathcal{R}^D$, representing the N elements (dialects), such that

$$\|x_i - x_j\| \approx L_{ij}, \quad \forall i, j \in \mathcal{N}, \tag{3.7}$$

where $\|...\|$ represents the norm of the vector—one can use the Euclidean distance metric. Notice that the x_i are not unique, so that MDS can be considered as the optimization problem of finding the N vectors $(x_1, ..., x_N)$ that minimize the function

$$F(x_1, ..., x_N) = \sum_{i<j} (\|x_i - x_j\| - d_{ij})^2. \tag{3.8}$$

Figure 3.12 represents the two-dimensional MDS plot of the 12 Mazatec dialects obtained from the same database and the Levenshtein distance matrix.[4] The MDS results are consistent with the unexpected observations discussed here: TE is far from HU and CQ and closer to AY. The MDS representation is more congruent with standard taxonomy of Mazatec dialects. Again, the dialect LO is far from all the other dialects.

3.6 Basque Regional Variation as a Case Study

Basque is a language isolate spoken in an area referred to as the 'Basque Country', located between north Spain and France. In this section, we review an investigation of the regional variation of the Basque language, based on the multi-modeling of the dynamical social aggregates of Basque (Léonard et al., 2015). The goal of the investigation is to reveal the emerging dialect networks, starting from geographic and social aggregates at different scales, obtained from the data present in the EAS database.[5] This section summarizes some of the results obtained by Léonard et al. (2015, 2012); and Léonard (2007).

3.6.1 Database and method

The data used are taken from the EAS database. The EAS data were collected through a survey involving elderly Basque speakers living in the places shown in Figure 3.13, which is a map

---

[4] The inbuilt MDS function of MATLAB was used.
[5] Euskararen Atlas Soziolinguistikoa, *Sociolinguistic Atlas of the Basque Language*, see Unamuno et al. (2012).

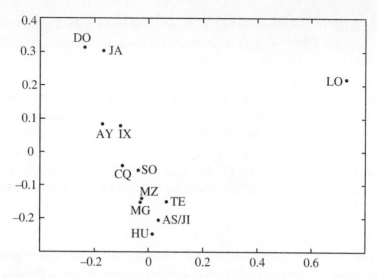

Figure 3.12 Two-dimensional projection from multidimensional scaling analysis (in linguistic space) of nouns in Kirk's data. (Adapted from Léonard et al. [2017].)

of the Basque Country (94 locations overall). The survey was carried out by the team of EAS of *Universidad del País Vasco*. During the survey, Basque speakers were interviewed and the answers were recorded and then digitalized as strings, referred to as a_{ik}. The strings were arranged in a matrix $A = \{a_{ik}\}$. The index k denotes the type of the questionnaire and runs from $k = 1$ to the total number of questionnaires $k = N_T$; the index i represents the location, where the interview was carried out, and runs from $i = 1$ to the total number of locations $i = N_L = 94$ (see Léonard et al. [2015] for further details). Two different methods were used in the original analysis and in the dialect clustering visualization.

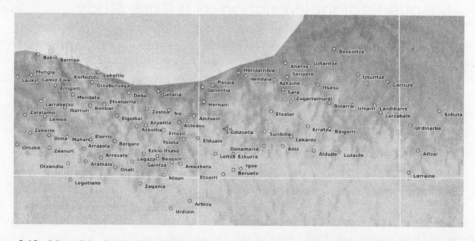

Figure 3.13 Map of the 94 localities surveyed by the EAS project, Unamuno et al. (2012). (Marble Virtual Globe: A World Atlas. KDE Platform [http://edu.kde.org/marble].)

- Weighted or non-weighted average Levenshtein distances between dialects.[6]
- Standard cladistics methods.[7]

In our discussion, we focus on the first method.

3.6.2 Morpho-(phono-)logical rules

In this section, we will illustrate the criteria used for computing weighted or unweighted (dis)similarities between dialects.

In the analysis of the Mazatec language presented in Section 3.5, the average Levenshtein distances, in this section referred to as L_{ij}, between dialects i and j was computed as the arithmetic average, Eq. (3.2), of all the contributions $L_{ij}^{(k)} = L(a_i^{(k)}, a_j^{(k)})$ measuring the distance between the two versions $a_i^{(k)}$ and $a_j^{(k)}$ of the single item $a^{(k)}$. This distance represents, for example, the difference between two versions of the same word in dialects i and j. In addition, we use the *weighted* average, Eq. (3.3), of the Levenshtein distance. The latter average is obtained by assigning a suitable statistical weight p_k to the generic contribution $L(a_i^{(k)}, a_j^{(k)})$. In our study, such weights are assigned according to a declarative model for morphophonological analysis that takes into account the type of change or preservation processes observed in the comparison between the different dialects and the reference dialect (Euskara batua). Such processes, the 'Structure', 'Feed', 'Void', and 'Loop' processes, are summarized in Table 3.5 (2nd column), together with their abbreviations S, F, V, and L (1st column) and the method is referred to as the *SFVL model*.

Table 3.5 Main operations according to the SFVL method.

Symbol	Process	Result	Explanation
S	Preserve STRUCTURE	Conservation/initial state	No change (standard Basque maintained)
F	FEED	Licensing/enrichment	Enrichment of the inner structure of phonological or morphological units
V	VOID	Government	Weakening or deletion of elements
L	LOOP	Reanalysis	Constituent inversion, for example, metathesis

To show how the SFVL coding works in practice, Table 3.6 illustrates its use in coding the structural change in the standard Basque *naiz* ('I am') and *zarete* ('you are'). The change processes (see the 'code' column of their abbreviations) are associated with a type (denoted by the letter A, B, C ..., see column 'type') that specifies the type of process in greater detail.

---

[6] See Heeringa (2004); Beijering et al. (2008); Brown et al. (2008). Other resources can be found online at: https://www.gabmap.nl/app/doc/manual/references1.html. https://github.com/coltekin/Gabmap/blob/master/doc/. flowcharts/levenshtein-tokenized.eps.

[7] For details concerning the cladistic analysis, see Brun-Trigaud et al. (2011); Corvaglia-Gaillard (2012).

Table 3.6 Examples of structural change correlates in the SFVL model: naiz and zarete.

EAS No.	[na-iz] 1Ab.Pres-Lx	Type	Parameters	Code	Weight	Combined weight	Σ
330	naiz	A	PRESERVE X	S	0		0
118	nais	B	FEED stridency	F	1		1
119	naix	B'	B & FEED stridency_palatal	F & F	1	1	2
133	naz	C	VOID root_nucleus	V	2		2
703	niz	D	VOID prefix_nucleus	V	2		2
134	nas	E	C & FEED stridency	F & V	4	1	5
120	nai	F	VOID outer_coda	V	4		4
136	na	G	VOID VC_root	V	4		4

EAS No.	[za-re-te] 5Ab.Pres-Lx-Pl	Type	Parameters	Code	Weight	Combined weight	Σ
119	zarete	A	PRESERVE X	S	0		2
202	sarete	A'	FEED stridency	F	2		2
302	serate	B	LOOP inverse_melody_Px-root (*sarete>serate*)	L & F	8	1	9
306	zaete	C	VOID root_onset	V	4		4
310	zeate	D	C & inverse_melody_Px-root (*zatete>seate*)	V & L	4	4	8
326	tzéate	D & E	FEED Pfx_initial_onset	F & L	2	4	6
404	zaaté	F	FEED left_nucleus	V & F	4	2	6
408	zate	G	VOID root_CV	V & L	4	4	8
301	sare	H	VOID CV_Sfx	V & L	4	4	8
110	sarie	I	FEED stem_contour (*saree>sarie*)	F & L	2	4	6
402	zabie	J	FEED labial_epenthetic_hiatus (°*zarie>°saie>°sawie>sabie*)	F & L	2	4	6
506	zaizte	K	LOOP reanalysis -iz-	L & F	8	1	9
503	ziezte	L	LOOP inverted_contour_Pfx-root (*zaizte>ziezte*)	L & L	8	4	12
425	zizte	M	VOID & LOOP Pfx_nucleus	V & L	4	4	8

We use the SFVL model to investigate emerging norms. It relies on an adaptation of the social cognitive model of language change (Hruschka et al., 2009). The model consists of form–function reanalysis, form–meaning mapping, iterated learning, replicators, and exemplars. At both ends of a polarity between conservation and change, preservation of structures resorts to trivial exemplars. In contrast, loop results from form–function reanalysis. Forms available in both extremes function as exemplars and can change under the pressure of iterated learning. Those that spread from a socio-cognitive norm to another, through contact or imitation, are replicators. Feed and void change exemplarity in a drastic way: feeding processes enrich exemplars, voiding processes make them lighter; if *naiz* ('I am') goes unmodified, *naz* (type C), *niz* (type D), or *na* (type G) undergo VOID processes. These morphophonological changes remain at the brink of semantic shift or change; hence, they do not endanger form–meaning mapping. In a way, this holds for types B and B' (*nais* and *naix*) as well, showing variation in the stridency of the coda fricative, challenging even less form–meaning mapping than the previous processes of the VOID type. As far as the example of *zarete* is concerned, comparing *zarete* and *sarete* gives an S pattern (no or insignificant change). Indeed, the comparison of *zarete* and *serate* (type B) implies an inversion of the vowel chain (vocalic melody), amounting to a LOOP. Type C *zaete* describes a gap in the form–meaning mapping, that is, there is a VOID root_onset as a result of intervocalic rhotic deletion (*zarate* > *zaete*). The reanalysis through inverse_melody_Px-root looping can be iterated in other dialects, as can be seen with type D *zeate*, which incorporates type C, and is thus charaterized as C & Inverse_melody_Px-root (*zaete* > *seate*). In the following sections, we use all the answers listed in Table 3.6.

Table 3.7 summarizes the main correlates of the SFVL model, contrasting two trends: change and leveling.

Table 3.7 Structural change correlates according to the SFVL model.

Change		Leveling	
Input → Output	Output → Input	Input → Output	Output → Input
FEED	**VOID**	**TRACE**	**LOOP**
Modification	Neutralization	Positional resilience	Repair or reform (change stem)
Interaction	Deleting	Opacity	Transparency
Pattern	Default	Equipollence	Regularity
Markedness	Inner structure skipping	Flip flop (iconic semiosis)	Analogy

3.6.3 Dialect network from the Levenshtein distances

In this section, the Levenshtein distances obtained from the analysis of the database are used to study and visualize the Basque dialect network.

Non-weighted Levenshtein distance-based network

Let us first consider the dialect networks obtained from the standard (non-weighted) Levenshtein distances L_{ij}. Each L_{ij} is the average of the Levenshtein distances between dialects

i and j, computed, using the whole corpus of auxiliary inflection, through Eq. (3.2) as the mean distances corresponding to the answers to the various questionnaires carried out in locations i and j. The resulting dialect network is shown in Figure 3.14 for (non-normalized) thresholds

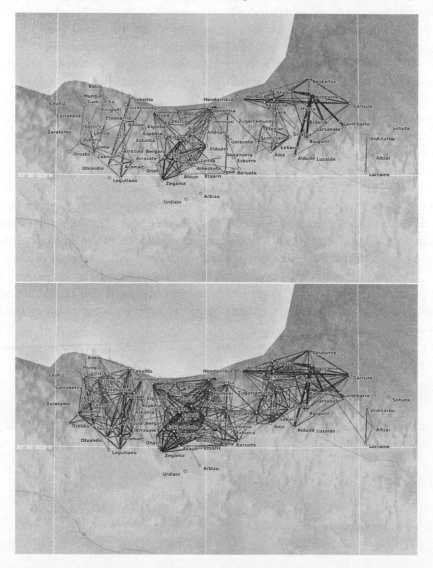

Figure 3.14 Non-weighted dialect network from the (arithmetic) average Levenshtein distances L_{ij}, using an upper threshold $L_{ij} < 1.8$ (top panel) and $L_{ij} < 2.0$ (bottom panel). (Marble Virtual Globe: A World Atlas. KDE Platform [http://edu.kde.org/marble].)

$d = 1.8$ (top panel) and $d = 2$ (bottom panel). At the threshold of differentiation, $d = 1.8$, a threefold division already appears, breaking up the network into the Bizkaian Western, the Central, and the Navarrese–Lapurdian Eastern dialect; the Zuberoan dialect also appears as a vertical isolated (Urdinarbe–Larraine) connection. The Central dialect shows a complex internal structure.

Table 3.8 An SFVL-based grid for ponderating the Levenshtein distance. Notice the basic hierarchy S≪F≪V≪L.

Process	Abbreviation	Weight (W)	Combined weight (CW)
PRESERVE X	S	0	none
FEED	F	2	1
VOID	V	4	2
LOOP	L	8	4
Synthetic	Y	3	1,5
ZUKA	Z	0,5	none
HIKA	H	0,5	none

At a threshold $d = 2$, the network in the bottom panel of Figure 3.14 looks intricate, but confirms some strong trends and shows some new clear features which include the following:

- Division of the Basque dialectal network into three subnetworks.
- The Eastern Souletin dialect of Zuberoa is now linked to the Navarrese–Lapurdian dialect.
- Presence of an inner structures for the three main dialects.
- Some locations, at strategic positions between the main clusters, act as effective crossroads for the spread of linguistic innovations.

Weighted Levenshtein distance-based network: SFVL method

In order to quantify the impact of innovation on the basis of their structural weight within the pool of features (Mufwene, 2001, 2013), one can attempt to construct the dialect network based on the mean Levenshtein distance between dialects by suitably weighting the different types of change with respect to the reference dialect, Euskara batua. The different weights used are listed in Table 3.8. The first four types of change, in order of increasing weight, are PRESERVE X, corresponding actually to the preservation of the structure and therefore with a corresponding zero weight, $W = 0$; FEED (F), with $W = 2$; VOID (V), with $W = 4$; and LOOP (L), with $W = 8$.

Furthermore, three factors of bias are weighted to reduce their incidence: Y for synthetic forms, with $W = 3$,[8] and the allocative ZUKA and HIKA forms, with $W = 0.5$. Since some questionnaires show combined types of changes, the additional column *Combined weight (CW)* allows one to take into account such combinations by summing the values in column *Weight (W)*, which represents the first process, to those of that which can be summed to the *Combined weight (CW)* column, representing a further process. Table 3.6 shows samples of ponderations for *naiz* 'I am' and *zarete* 'you are' (last three columns).

---

[8] Variable V 147 (90310) edin [DADIN]: 'We gave him some money so that he could go on holidays', in 108 yoáteko, 109, joatéko, 120 dxúteko (synthetic forms) vs. 110 yoan deiten, 116 yoan daitén, etc. (analytical forms).

The analysis was limited to 12 variables,[9] in view of the comparison between the results of cladistics and dialectometry, corresponding to the analysis of 1 167 tokens (instead of the 3 816 available from the aditza corpus of the EAS survey). The SFVL-weighted dialect networks obtained for the thresholds $d = 1.6$ and $d = 2$ are shown in Figure 3.15. Not only are these

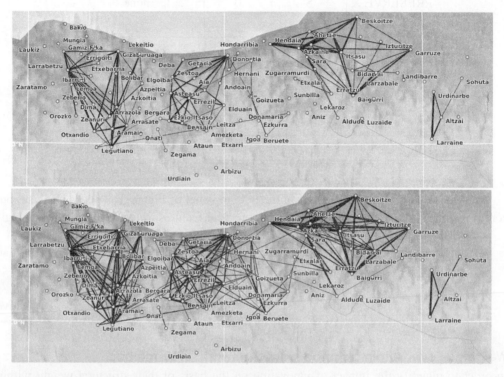

Figure 3.15 Weighted network of Basque dialects from the pondered average Levenshtein distances $L_{ij}^{(w)}$, using an upper threshold $L_{ij} < d$, with $d = 1.6$ (top panel) and $d = 2$ (bottom panel). See text for details. (Marble Virtual Globe: A World Atlas. KDE Platform [http://edu.kde.org/marble].)

results congruent with the results obtained earlier for the unweighted networks, they actually highlight even more clearly the subclusters and transitional areas formed.

An interesting link in the southern region connects Legutiano in the West to Donamaria, suggesting a possible southern track for the spreading of innovations from Southern Bizkaia to Southern Navarra. Since these links characterize the systems more in ontogenetic terms through flows of innovations and structural entropy than about phylogenetic terms, they deserve further studies.

---

[9] V144 (90010) izan [NAIZ]: 'I am young', V145 (90060) izan [ZARETE]: 'You too are young', V146 (90110) izan [GINEN]: 'We too were young a long time ago', V147 (90310) edin [DADIN]: 'We gave him some money so that he could go on holidays', V148(90360) edin [DAITEKE]: 'Your friend can come to my place', V149(90410) izan [ZAIT]: 'I like this meat', V150 (90520) izan [ZITZAIDAN]: 'Before, I used to like eating meat, too', V151 (90590) izan [ZITZAIZKION]: 'Before, my father used to like strawberries, too', V152 (90700) edun [DUT]: 'Today I didn't eat much', V153 (90730) edun [DU]: 'He didn't eat a lot', V154 (90750) edun [DUZUE]: 'You too, you didn't eat a lot', V155 (90760) edun [DUTE]: 'My friends have eaten a sandwich'.

3.6.4 Concluding remarks on the Basque analysis

The picture emerging from the processing of the EAS corpus points at a threefold division of Basque dialects, possibly with further subdivisions. Between these blocks and sub-blocks, some intermediate 'default areas' can be identified, which must have played an important role in the history of the language, propagating features between emerging blocks, or main dialect clusters, that is, Bizkaian, Gipuzkoan, Navarrese–Lapurdian. The intermediate zones draw clear frontiers within the three main blocks and, not coincidentally, reflect the underlying topology of the diffusion routes.

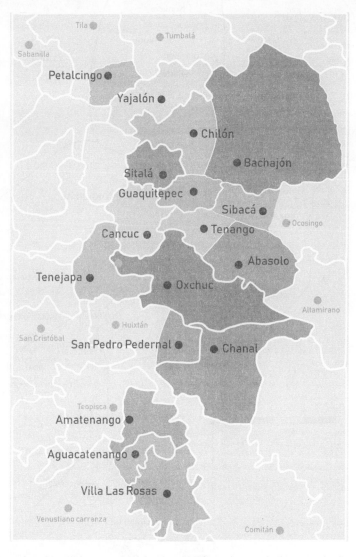

Figure 3.16 Map of the localities surveyed in the ALTO project and discussed in Section 3.7. (From http://alisto.org.)

3.7 Case Study: The Tseltal Diasystem

In this section, the case study of the Tseltal language, a Western Mayan language of the highlands of Chiapas, in Southern Mexico, is discussed. Different types of networks are

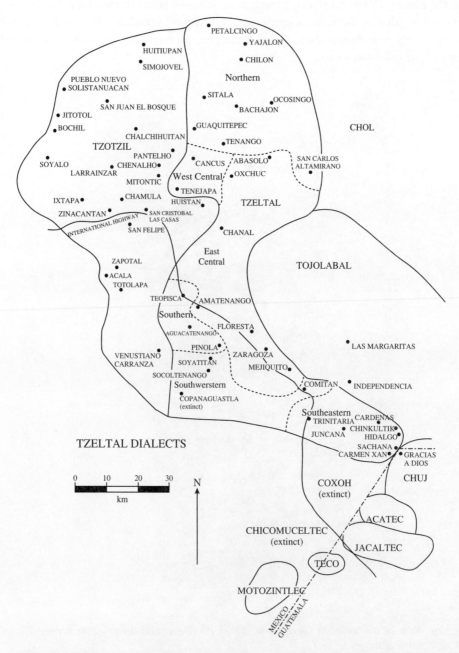

Figure 3.17 Original map of the dialect areas, the Greater Tseltalan (from Campbell [1987]).

considered in order to characterize the relations between Tseltal dialects. The networks were built using different quantities and focusing on different aspects, namely (a) morphology, (b) phonology, and (c) lexicology. The analysis illustrated in this section is detailed in Polian et al. (2014) and was based on the ALTO database.[10]

3.7.1 Subdivision of the Tseltal Dialects

The standard classification of Tseltal dialects generally retains a subdivision into three parts: Northern, Central, and Southern Tseltal (see Table 3.9 and the map in Figure 3.16, showing the dialects considered in the analysis). However, the general picture contains much more

Table 3.9 Tseltal dialects of the analysis and respective abbreviations.

		Northern 1			**Northern 2**
1.	PE	Petalcingo	6.	GU	Guaquitepec
2.	YA	Yajalón	7.	SB	Sibacá
3.	CHI	Chilón	8.	TG	Tenango
4.	BA	Bachajón			
5.	ST	Sitalá			
		West Central			**East Central**
12.	TP	Tenejapa	9.	CA	Cancuc
			10.	AB	Abasolo
			11.	OX	Oxchuc
			13.	SP	San Pedro Pedernal
			14.	CHA	Chanal
		Southern			
15.	AM	Amatenango			
16.	AG	Aguacatenango			
17.	VLR	Villa Las Rosas			

details, as illustrated by the original map by Campbell (1987), shown in Figure 3.17, of the Greater Tseltalan: the Central Tseltal can be divided into East Central and an autonomous West Central sub-dialect (Tenejapa), a hypothesis which is sustained also by the dialectometric analysis discussed here, while the homogeneity of OX, CHA, and AB varieties is generally recognized. The intricacy of the Southern Tseltal is shown in the map by the subdialect chain extending in the Southern area, which further complexifies into South Western and South Eastern Tseltal, as opposed to Southern Tseltal (Aguacatenango) (see also Hopkins [1970] and Kaufman [1972]). In addition, Hopkins suggests a division of the Northern dialect into

---

[10] The data processed in this section have been gathered, since 2008, through fieldwork within the ALTO project (*Atlas Lingüístico del Tseltal Occidental*). A comprehensive list of all the variables is available on the project web page: http://alisto.org. ALTO aims at providing a first hand survey of the dialectal variation in Tseltal. The project contains 70 structural variables divided among lexicon (36), phonology (13), and morphfology (21).

two dialect chains, (PE, YA, CHI, BA, ST) and (GU, TG, SB), and two Eastern Tseltal varieties (Ocosingo and Altamirano, in grey in Figure 3.16). Comparative data with other Mayan languages are available in Kaufman and Justeson (2003).

Notice that the map of Campbell shows Tseltal within the context of its dialect continuum with another important language of the area, at the left of the map — *Tsotsil* — to a large extent, mutually intelligible with Tseltal. It also enumerates other kin languages in the South such as Chuj, Acatec, Jacaltec (all from the Q'anjobalian subfamily of Western Mayan languages), and some extinct languages, such as Coxoh and Chicomuceltec. The two southernmost subdialects at the bottom of the Tseltal area (South Western and South Eastern) are now extinct.

3.7.2 Analysis of the Tseltal database: Morphology

The integration of morphosyntactic data into dialect classifications according to typological criteria (Kortmann, 2004) is an emerging trend in modern dialectology since it provides powerful contrasts as compared to more classical approaches. This is because the taxonomies are based mainly on phonological or lexical variables.

Phonological variation is more trivial than morphosyntactic variation and more embedded in the specific typology of the sound system of a given language. On the other hand, morphosyntactic variation is more relevant for the understanding of universal trends (universal grammar), although it depends on core (or local) grammar specificities. For example, just like the other Mayan languages, Tseltal has *ergative alignment*, that is, it differentiates between the agent of a transitive verb (marked with possessive/ergative prefixes) and the subject of an intransitive verb or the patient or object of a transitive verb (marked with absolutive suffixes).

In order to gain a realistic idea of the complexity and diversity of the Tseltal dialects, a sample of 21 morphosyntactic variables processes is used, which are listed in Table 3.10.

Such variables have been indexed in Table 3.11 for all the 17 localities in Table 3.9, where each letter stands for one of the types defined in Table 3.10: most recurrent types are written in capital fonts, whereas more endemic types appear in small fonts; types inside parentheses are secondary and can be considered as belonging to the realm of weak polymorphism.

The variables' indexation in Table 3.11 is used to compare the different languages by computing a *similarity index* s_{ij} between pairs of dialects i and j (rather than the Levenshtein distance, see Goebl [1981]; Geoebl [2003]).[11] The similarity index s_{ij}, for each pair of dialects i and j ($i, j = 1, \ldots, 17$ in Table 3.9), is computed by summing a contribution $M_k(i, j)$ for each variable k, obtained by comparing the values in Table 3.11 of dialect i and j,

$$s_{ij} = \sum_k M_k(i, j). \tag{3.9}$$

Each match between the values of the kth variable in dialects i and j add a contribution $M_k(i, j) = +1$ to the similarity index s_{ij}; whereas any mismatch is valued 0. Furthermore, there can be some further rescaling due to statistical or linguistic reasons.

---

[11] A description of dialectometric methods can be found online at www.dialectometry.com/ and www.let.rug.nl/kleiweg/L04/.

Table 3.10 Morphosyntactic variables.

k	Variable	Explanation
1	1ERG	Ergative 1 singular subject agreement marking: (a) *k-* vs. (b) *j-*
2	ABS+PL	Order of affixation between the absolutive suffix and the plural marker: (a) ABS-PL vs. (b) PL-ABS
3	2ABS.PL	Second person plural absolutive marker: (A) *-ex* vs. (b) *-atik* or the hybrid form (c) *-atex*
4	EXCL	First person plural exclusive: (A) *-yotik* vs. (B) *-jo'tik*, or (C) *-kotik*, or (D) *-tikon*, or (e) none.
5	APL+1ABS	Applicative suffix combined to first person singular absolutive: (a) *-bon* vs. (b) *-ben*
6	APL+2ABS	Applicative suffix combined to second person singular absolutive: (A) *-beyat* vs. (b) *-bat* or (c)) *-bet*
7	APL+PAS	Applicative suffix combined to passive marker: (a) *-bot* vs. (b) *-bet*
8	VI PROHIB	Prohibitive form of intransitive verbs: (a) incompletive vs. (b) imperative
9	ITER	Iterative suffix: (A) *-ilan* / *-ulan* vs. (b) *-ila(y)* / *-ula(y)*
10	DISTR	Distributive suffix: (A) *-tik* + ITER vs. (b) *-ti* + ITER or (c)) *-ta* + ITER
11	PROG.O3	Progressive construction for transitive verbs with third person object: (a) Aux. + V. inaspectual vs. (b) Aux. + ABS + V. infinitive or (c)) Aux. + V inflected
12	PROG.O1/2	Progressive construction for transitive verbs with first or second person object: (a) Aux. + V. inaspectual vs. (b) Aux. + ABS + V. inaspectual or (c)) Aux. + V. inflected, or (d) Aux. + ABS + V. possessed nominalized stem
13	PROG AUX	Auxiliar for progressive aspect: (a) *yakal, yak* vs. (b) *nok'ol, no'*, or (c)) *yipal*, or (d) *yak'al*
14	INC+2ERG	Reanalysis of the *k* component of incompletive auxiliar *yak* as a prefix before the *a*-prefix of second person subject agreement: (A) no reanalysis vs. (b) reanalysis
15	INC+3ERG	Reanalysis of the *k* component of incompletive auxiliar *yak* as a prefix before the *a*-prefix of third person subject agreement: (A) no reanalysis vs. (b) reanalysis
16	DIR 'come'	Afferent (from there to here) directional 'coming': (a) *tal* vs. (b) *tel* or (c)) *tael*
17	DIR 'go'	Efferent (from here to there) directional 'going away': (a) *bahel, bael* vs. (b) *beel, bel*, or (c)) *k'ajel, k'ael, k'al*
18	INVERSE	Inverse voice (suffice *-on*): (A) maintained vs. (b) neutralized for passive
19	CVL AGENT	Construction of agentive light verb: (A) none vs. (b) occurs
20	PAS NF	Reanalysis of transitive infinitive forms as finite and passive: (A) none vs. (b) occurs
21	PART INT	Interrogative particle (polarity questions): (a) *bal* (enclitic of 2nd possessive) vs. (b) *me* (phrase initial particle)

- *There are more than a single value corresponding to a given variable in a given location.* In general, there can be more values $v(k)$ of the same variable k co-occurring in the same location, so that the sum, Eq. (3.9) has to be understood to run over all the values $v(k)$ of each variable k,

$$s_{ij} = \sum_{k} \sum_{v(k)} M_{v(k)}(i,j), \qquad (3.10)$$

where the factor M, which used to be equal to 1 in case of match and to 0 in case of mismatch, is now a number in the interval $M \in (0, 1)$. For example, if there are two possible values a, b of the same variable in a certain location i, they are assigned the

Table 3.11 Indexation of the 21 morphosyntactic variables.

k	Acronym	W_k	PE	YA	CHI	BA	ST	GU	SB	TG	CA	AB	OX	TP	SP	CHA	AM	AG	VR
1	1ERG	5%	a	b	b	b	b	b	b	b	b	b	b	b	b	b	b	b	b
2	ABS+PL	5%	a	a	a	a	a	a	a	a	a	a	b	a	a	b	a	a	a
3	2ABS.PL	5%	A	A	A	A	A	A	A,b	A	A,b	A,b	A	A,b	b	A	A	c	
4	EXCL	8%	A	B	C	C	B	B	D	e	B,e	e	e	B	e	e	e	e	D
5	APL+1ABS	6%	a	a	a	a	a	a	a	a	a	a	b	a	a	a	a	a	
6	APL+2ABS	8%	b	b	b	b	b	b	b	c	c	b	c	c	b	A	b	c	
7	APL+PAS	4%	a	a	a	a	a	a	a	a	a	a	b	a	a	a	a	a	
8	VI PROHIB	5%	a,b	a	a	a	a	a	a	a	a	a	a	a	a	b	a,b	b	
9	ITER	2%	b	b	A	A	A	A	A	A,b	b	b	b	A	b	A	A	A	
10	DISTR	2%	A	A	A	A	A	A	A	b	b	b	A	A	b	c	c	c	
11	PROG.O3	8%	a	a	a	a	a	a	a,b	a	a,b	b	b	b,(a)	b,c	b	b,c	b	b,(c)
12	PROG.O1/2	6%	a	a	a	a	a	a	a	a	a,b	a,b	a,b,(d)	c	a,b	c	d	d,(c)	
13	PROG AUX	4%	a	a	a	a,b	a	a	a	a	a	a	a	a	a	c	a,c	a,(d)	
14	INC+2ERG	4%	b	b	A,b	b	A	A,b	A,(b)	A	A	A	A	A	A	A	b	A	
15	INC+3ERG	2%	b	A	A	A	A	A	A	A	A	A	A	A	A	A	A	A	
16	DIR venir	3%	a	a	a	a	a	a	a	b	b	b	a	a,c	b	a	a	a	
17	DIR ir	3%	a	a	a	a	a	a	a	b	b	b	b	a,c	b	a,c	d	a,d,(c)	
18	INVERSO	4%	b	b	b	b	b	b	b	b	b	b	b	b	b	b	b	A	
19	CVL AGENT	4%	b	b	b	b	b	b	b	b	b	b	b	b	b	A	A	A	
20	PAS NF	4%	A	A	A	A	A	A	A	A	A	A	A	A	A	b	b	b	
21	PART INT	8%	a	a	a	a	a	a	a	a	a	a	a	a	a	b	b	b	
	TOTAL	100%																	

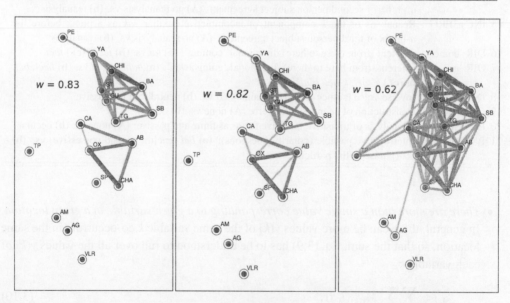

Figure 3.18 Maps obtained from the morphological similarity matrix, Eq. (3.12), for three different values of the threshold w [normalized in the interval (0,1)].

weights $p_a(i) = 1/2$ and $p_b(i) = 1/2$; if there are three values a, b, c, they are assigned the weights $p_a(i) = 1/3$, $p_b(i) = 1/3$, $p_c(i) = 1/3$. If one of the attribute is secondary,

Table 3.12 Similarity index from morphosyntactic data.

	PE	YA	CHI	BA	ST	GU	SB	TG	CA	AB	OX	TP	SP	CHA	AM	AG	VLR
PE	–	83.1	79.6	79.6	77.1	79.6	75.9	76.5	60.1	51.0	59.0	46.1	51.0	55.9	28.8	50.5	26.1
YA	83.1	–	88.5	88.5	94.0	96.5	84.8	85.4	74.0	59.9	67.9	62.9	59.9	64.8	32.6	55.6	30.0
CHI	79.6	88.5	–	97.0	90.5	92.0	88.5	87.9	69.5	60.4	68.4	55.4	64.4	65.3	37.1	56.1	34.5
BA	79.6	88.5	97.0	–	86.5	89.0	85.3	83.9	65.5	56.4	64.4	51.4	60.4	61.3	34.6	57.1	31.0
ST	77.1	94.0	90.5	86.5	–	98.5	88.3	89.4	76.0	61.9	69.9	64.9	65.9	66.8	38.6	53.6	36.0
GU	79.6	96.5	92.0	89.0	98.5	–	88.5	87.9	74.5	60.4	68.4	63.4	64.4	65.3	37.1	56.1	34.5
SB	75.9	84.8	88.5	85.3	88.3	88.5	–	85.6	73.3	66.1	74.1	59.7	69.1	71.0	41.9	60.4	47.7
TG	76.5	85.4	87.9	83.9	89.4	87.9	85.6	–	75.4	73.0	81.0	56.3	75.0	79.1	44.0	59.0	35.2
CA	60.1	74.0	69.5	65.5	76.0	74.5	73.3	75.4	–	89.9	81.9	73.4	75.1	78.8	40.8	50.6	41.8
AB	51.0	59.9	60.4	56.4	61.9	60.4	66.1	73.0	89.9	–	92.0	68.0	81.0	90.1	42.9	54.8	43.8
OX	59.0	67.9	68.4	64.4	69.9	68.4	74.1	81.0	81.9	92.0	–	60.0	73.0	98.1	42.9	62.8	35.8
TP	46.1	62.9	55.4	51.4	64.9	63.4	59.7	56.3	73.4	68.0	60.0	–	59.5	56.9	34.3	37.3	33.0
SP	51.0	59.9	64.4	60.4	65.9	64.4	69.1	75.0	75.1	81.0	73.0	59.5	–	71.1	58.8	55.6	52.8
CHA	55.9	64.8	65.3	61.3	66.8	65.3	71.0	79.1	78.8	90.1	98.1	56.9	71.1	–	39.8	59.6	35.8
AM	28.8	32.6	37.1	34.6	38.6	37.1	41.9	44.0	40.8	42.9	42.9	34.3	58.8	39.8	–	62.9	59.9
AG	50.5	55.6	56.1	57.1	53.6	56.1	60.4	59.0	50.6	54.8	62.8	37.3	55.6	59.6	62.9	–	59.5
VLR	26.1	30.0	34.5	31.0	36.0	34.5	47.7	35.2	41.8	43.8	35.8	33.0	52.8	35.8	59.9	59.5	–

the secondary attribute is given a weight lower than the others: for example, if the variable k can assume the two attributes $a,(b)$ in location i, the first value is given a $p_a(i) = 3$, $p_{(b)}(i) = 3/4$ and the secondary value (written in parentheses) a $p_{(b)}(i) = 1/4$. If there are three attributes, for example, a, b, (c), the conventional assignments are $p_a(i) = p_b(i) = 3$, $p_{(c)}(i) = 3/7$, and $p_c = 1/7$, respectively. However, if such a situation arises, the same attribute $v(k)$ may have been assigned different weights in the two dialects i and j, for example, $p_a(i) \neq p_a(j)$, and choosing the contribution $M_{v(k)}(i, j)$ in Eq. (3.10) becomes a problem. A possibility is to choose the smaller weight between the two weights (referred to as the 'intersection of shared properties'), that is, $M_{v(k)}(i, j) = \min(p_a(i), p_a(j))$. Alternatively, one can choose an average of the two weights ('shared properties'), $M_{v(k)}(i, j) = [p_a(i) + p_a(j)]/2$. The similarity indexes used in the analysis are computed using two methods (see Polian et al. [2014] for details).

- Linguistic relevance. An additional rescaling of the contribution $M_{v(k)}(i, j)$ in Eq. (3.9) comes from a weighting procedure, with weights W_k, which expresses the relative linguistic importance of the respective variables. The weights W_k are listed in the third column of Table 3.11 (expressed in percentage), so that the final formula used to compute the similarity index is

$$s_{ij} = \sum_k W_k \sum_{v(k)} \bar{M}_{v(k)}(i, j). \tag{3.11}$$

The results for the similarity index of the Tseltal dialects, computed from Table 3.11, are listed in Table 3.12 as percentages. To make an effective representation of these data, one can introduce a threshold w and plot the network obtained considering only the links between dialects such that the corresponding similarity index is

$$s_{ij} > w. \tag{3.12}$$

This network selects the most strongly connected dialects, removing weaker links with $s_{ij} \leq w$. Rather than working in an abstract space, it is instructive to embed the network in the actual geographical map of the Tseltal area, in which the geographical locations coincide with the corresponding dialects. This provides interesting additional information about the relations between dialects. To enhance visualization of the different link strengths, the larger is the value of s_{ij}, the thicker and darker is the link drawn. The results are shown in Figure 3.18 for three different values of the threshold w, rescaled in the interval $(0,1)$ (that is using a similarity index rescaled as $s_{ij} \rightarrow s_{ij}/s_{max}$, where $s_{max} = \max\{s_{ij}\}$ is the largest value of the similarity matrix). The three value selected, $w = 0.82, 0.83, 0.62$, highlight some of the significant structures in the network, by providing a clear picture of communal aggregates; as the most sensitive to discourse, morphosyntax lies at the core of the discursive and pragmatic dimension of language. The higher the score, the more similar or convergent the grammar of the clusters in the graphs; the lower the score, the more loose the interrelation between the aggregates. At the highest threshold, $w = 0.83$, the graph points out two distinct choremes (or geolinguistic nuclei), one in the center (OX & CHA & CA, AB) and the other in the north (YA & ST, CHI & BA & others), and some isolated vertices in the south. The slightly lower value, $w = 0.82$, forms the first bridge between the major north and central families. At the lowest value shown, $w = 0.62$, the first connections involving some southern dialects are finally formed; but at such a w value, basically all the other nodes in the center and in the north have already been merged in an almost fully connected network, including the diversified TP dialect: at this threshold, TP connects as much with the center (CA & OX, as could be expected) as with the north (ST & YA). These facts, together with the persistence of the isolation of the southern vertices at such a low threshold, suggest an intricate character and a string internal heterogeneity of the southern dialect. In sharp contrast with canonical classifications, the southern dialect does not seem to exist anymore, appearing fragmented as a cluster of isolated varieties.

We can observe how, in all this, the dialect of OX (Oxchuk, 33 780 inhabitants), turns out to be a chaining dialect or, metaphorically, a broker, since, as the threshold is lowered more and more, it is both the first location to connect central and northern dialects, and the first one to link southern dialects to the other dialects. Not coincidentally, it is a very dynamic crossroad—in fact, its name means 'three nodes' from *ox(eb)* ('three') and *chuk* ('node', 'tie'), alluding to the confluence of three important roads across the Central Chiapas Highlands. Places, such as Carhaix in France and Oxchuk in Mexico, at the center of regional road networks, are typically strategic market places that effectively catalyze features originating from their neighborhood.

3.7.3 Analysis of the Tseltal database: Phonology

The networks corresponding to the phonological features look quite different from the morphological one, although they have some consistent trends. Table 3.13 displays the set of 13 phonological variables relevant to the Western Tseltal dialect network. Table 3.14 lists

Table 3.13 Phonological variables.

k	Variable	W_k	Explanation
1	p' – b	14%	Opposition of an ejective labial /p'/ vs. a simple unvoiced labial stop /p/ (an idiosyncrasy inherited from Proto-Tseltal, found in Petalcingo only, in the peripheral north of the domain)
2	x – h	12%	Patterns of the laringovelar correlation /x/ (velar fricative, with j notation as in Spanish) vs. /h/ (laryngeal fricative) inherited from Proto-Tseltal according to dialects (see Gendrot et al. 2010, Léonard et al. 2013)
3	h + Sonorant	7%	Variation (or deletion) of the laryngeal fricative before a sonorant
4	h + Ejective	5%	Variation (or deletion) of the laryngeal fricative before an ejective
5	CV*h*CVC	5%	Variation (or deletion) of the laryngeal fricative at word-level in a template as CV*h*CVC
6	V*h*V	9%	Interrupted vowels/complex nuclei [Léonard et al. (2011), Brown and Wichmann (2004)]
7	V'V	6%	Interrupted vowels/complex nuclei [Léonard et al. (2011), Brown and Wichmann (2004)]
8	V'C	6%	Preconsonantal laryngeals
9	h + Stop/Affricate	7%	Preconsonantal laryngeals
10	s > <š>	8%	Sibilant palatalization through assimilation
11	VV > V	8%	Long vowels reduction
12	VV / VjV	5%	Diatopic variation between long and complex nuclei
13	CV*h*-C...	8%	Laryngeal fricative before a consonantal suffix
	TOTAL	100%	

the similarity indexes [obtained from the phonological variables in Table 3.13 and normalized as percentages] between Tseltal dialects. As most of the phonological variables listed in the table happen to be fine-grained details linked with glottic features (ejectives, complex nuclei) or glottal harmony, syllabic compression in stems, or assimilations, the phenomenology is more sensitive to thresholds than in the morphological matrix. Interesting structures appear within the strict w threshold range between $w_2 = 0.97$ (close to total similarity) and $w_1 = 0.90$ (high similarity) as shown in Figure 3.19—the w s are normalized in the interval (0,1). In the phonological dimension, Southern Tseltal does exist and shows up with the AM & AG chain, while VLR lingers as an isolate. Northern Tseltal hardly emerges in the highest thresholds; it eventually shows up as a node–line graph at $w = 0.90$. In all this, the central dialect is consistently embedded with in the Southern area, and the Western Central dialect TP shows a double link with the southern dialect chain (AM, AG, & VLR) to the south, and with GU (Guaquitepec) to the north, that is, to peripheral varieties. While the central and northern areas congregate, the northern subdiasystem lines up in a PE-YA-CHI-BA chain; ST seems to be no more than a satellite of YA. The remainder of the northern area clusters with the center, GU with TP, but also with the southern chain. Thus, an asymmetry with respect to the former morphological configuration, characterized by a center clustering with the north, appears striking, with the center clustering with the south and the peripheral satellite subdialects of the center.

Table 3.14 Pondered matrix for phonological similarity [matrix elements are normalized as percentages in the interval (0,100)].

	PE	YA	CHI	BA	ST	GU	SB	TG	CA	AB	OX	TP	SP	CHA	AM	AG	VLR
PE	–	64.6	78.3	95.0	62.8	56.8	72.6	59.0	57.5	62.9	18.0	51.0	48.4	18.0	54.5	56.0	60.0
YA	64.6	–	78.9	59.6	93.1	79.4	82.0	70.6	69.1	72.9	29.6	78.6	60.0	29.6	74.1	75.6	70.6
CHI	78.3	78.9	–	83.3	87.0	80.5	88.9	76.9	75.4	79.8	35.9	76.9	66.3	35.9	78.4	79.9	78.3
BA	95.0	59.6	83.3	–	67.8	61.8	77.6	64.0	62.5	67.9	23.0	56.0	53.4	23.0	59.5	61.0	65.0
ST	62.8	93.1	87.0	67.8	–	87.5	90.1	78.8	77.3	81.0	37.8	86.8	68.1	37.8	82.3	83.8	78.8
GU	56.8	79.4	80.5	61.8	87.5	–	81.4	85.4	86.9	88.3	49.4	93.4	79.8	49.4	91.9	90.4	88.8
SB	72.6	82.0	88.9	77.6	90.1	81.4	–	88.6	87.1	90.9	47.6	80.6	78.0	47.6	84.1	85.6	80.6
TG	59.0	70.6	76.9	64.0	78.8	85.4	88.6	–	98.5	98.3	59.0	92.0	89.4	59.0	95.5	97.0	83.0
CA	57.5	69.1	75.4	62.5	77.3	86.9	87.1	98.5	–	96.8	62.5	90.5	92.9	62.5	97.0	95.5	81.5
AB	62.9	72.9	79.8	67.9	81.0	88.3	90.9	98.3	96.8	–	57.3	90.3	87.7	57.3	93.8	95.3	86.9
OX	18.0	29.6	35.9	23.0	37.8	49.4	47.6	59.0	62.5	57.3	–	51.0	71.4	100.0	59.5	56.0	42.0
TP	51.0	78.6	76.9	56.0	86.8	93.4	80.6	92.0	90.5	90.3	51.0	–	81.4	51.0	95.5	97.0	83.0
SP	48.4	60.0	66.3	53.4	68.1	79.8	78.0	89.4	92.9	87.7	71.4	81.4	–	71.4	89.9	86.4	72.4
CHA	18.0	29.6	35.9	23.0	37.8	49.4	47.6	59.0	62.5	57.3	100.0	51.0	71.4	–	59.5	56.0	42.0
AM	54.5	74.1	78.4	59.5	82.3	91.9	84.1	95.5	97.0	93.8	59.5	95.5	89.9	59.5	–	98.5	84.5
AG	56.0	75.6	79.9	61.0	83.8	90.4	85.6	97.0	95.5	95.3	56.0	97.0	86.4	56.0	98.5	–	86.0
VLR	60.0	70.6	78.3	65.0	78.8	88.8	80.6	83.0	81.5	86.9	42.0	83.0	72.4	42.0	84.5	86.0	–

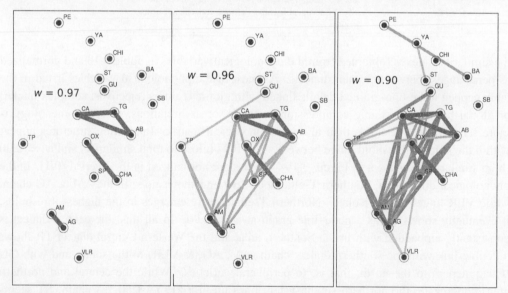

Figure 3.19 Networks of Tseltal varieties according to phonological clustering of pondered similarity index. The thresholds are normalized in the interval (0,1).

3.7.4 Analysis of the Tseltal database: Lexical variation

To test what the lexical variation provides, a lexicon most accountable for dialect variation in Western Tsetal was selected using the same similarity index as explained in the morphological and phonological analysis; the lexical variables are listed in Table 3.15. The lexicon items are pondered according to the weight of isoglosses. Table 3.16 lists the similarity indexes

Table 3.15 Pondered lexical variables.

k	Lexicon	W_k	k	Lexicon	W_k	k	Lexicon	W_k
1	what	6%	13	get wet	1%	25	a lot *bay{e/a}l*	4%
2	now	4%	14	first	4%	26	day, sun	4%
3	children	4%	15	cotton	1%	27	no	6%
4	village	3%	16	glanders	2%	28	far	2%
5	right	1%	17	bed	1%	29	talk	2%
6	blanket	1%	18	cricket	1%	30	see off	2%
7	(his) horn	1%	19	egg	3%	31	play	3%
8	tortilla	2%	20	foot	3%	32	pal, buddy	2%
9	can	4%	21	horse	3%	33	woman's eldest brother	1%
10	DEM.LOC	6%	22	ADMIR	3%	34	a great deal *mi*	2%
11	we go	3%	23	light up	1%	35	sky	2%
12	also	6%	24	evening	3%	36	high, long	3%
							TOTAL	100%

Table 3.16 Similarity indexes [normalized as percentages in the interval (0,100)] from the pondered lexical variables.

–	PE	YA	CHI	BA	ST	GU	SB	TG	CA	AB	OX	TP	SP	CHA	AM	AG	VLR
PE	–	93.0	96.9	95.6	91.4	87.5	84.0	56.5	62.8	32.6	26.0	53.0	29.0	25.3	46.5	30.9	38.0
YA	93.0	–	93.9	92.6	93.4	88.5	79.0	55.5	55.8	31.6	25.0	46.0	28.0	24.3	45.5	33.9	35.0
CHI	96.9	93.9	–	98.5	91.3	88.9	83.9	56.4	62.6	32.5	25.9	52.9	27.9	25.1	45.4	29.8	36.9
BA	95.6	92.6	98.5	–	89.0	87.1	81.6	56.1	62.4	33.3	26.6	52.6	29.6	25.9	46.1	30.9	37.6
ST	91.4	93.4	91.3	89.0	–	91.4	83.4	59.4	65.6	37.8	32.4	55.9	33.4	31.6	47.9	34.3	38.4
GU	87.5	88.5	88.9	87.1	91.4	–	85.5	68.5	70.3	43.4	39.5	57.5	40.5	38.8	49.0	36.4	34.5
SB	84.0	79.0	83.9	81.6	83.4	85.5	–	71.0	70.8	47.9	42.0	58.0	43.0	41.3	45.5	35.9	40.0
TG	56.5	55.5	56.4	56.1	59.4	68.5	71.0	–	81.8	70.4	66.5	67.5	67.5	65.8	44.0	39.4	38.5
CA	62.8	55.8	62.6	62.4	65.6	70.3	70.8	81.8	–	67.9	66.8	87.3	67.8	66.0	44.3	40.6	36.8
AB	32.6	31.6	32.5	33.3	37.8	43.4	47.9	70.4	67.9	–	95.6	72.1	92.6	94.9	39.1	41.0	37.6
OX	26.0	25.0	25.9	26.6	32.4	39.5	42.0	66.5	66.8	95.6	–	71.0	93.0	99.3	36.5	38.9	33.0
TP	53.0	46.0	52.9	52.6	55.9	57.5	58.0	67.5	87.3	72.1	71.0	–	76.0	72.3	51.5	50.1	44.0
SP	29.0	28.0	27.9	29.6	33.4	40.5	43.0	67.5	67.8	92.6	93.0	76.0	–	94.3	43.5	41.9	40.0
CHA	25.3	24.3	25.1	25.9	31.6	38.8	41.3	65.8	66.0	94.9	99.3	72.3	94.3	–	37.8	40.1	34.3
AM	46.5	45.5	45.4	46.1	47.9	49.0	45.5	44.0	44.3	39.1	36.5	51.5	43.5	37.8	–	81.4	80.0
AG	30.9	33.9	29.8	30.9	34.3	36.4	35.9	39.4	40.6	41.0	38.9	50.1	41.9	40.1	81.4	–	76.9
VLR	38.0	35.0	36.9	37.6	38.4	34.5	40.0	38.5	36.8	37.6	33.0	44.0	40.0	34.3	80.0	76.9	–

[obtained from the lexical variables in Table 3.15 and normalized in the interval (0,100)] between Tseltal dialects. Lexical differentiation is less fine-grained than phonological variation (think of the opposition between Classical Latin *caput* versus Imperial Latin *testa* in Romance languages); whereas phonological variation is far more complex. For these reasons, structures appear in a bolder threshold range here, within the w threshold interval $w = (0.75, 0.50)$ as shown in Figure 3.20 [thresholds normalized in $(0,1)$]. It is remarkable that the results for

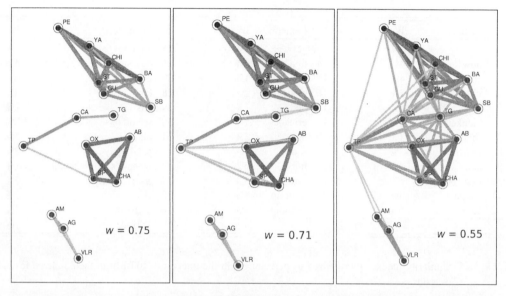

Figure 3.20 Networks of Tseltal varieties according to lexical clustering of pondered similarity index [here the threshold is normalized in the interval $(0,1)$].

$w = 0.75$ provide a picture akin to the canonical classification of the Hopkins, Kaufman, and Campbell model, with a clear-cut threefold division between the north, the center (with the TP subdialect making up a node–line subgraph with CA and TG) and the south. The central core OX–AB–SP–CHA emerges very clearly as a complete graph, with an impressive density of links, and at $w = 0.55$, the TP Central Western subdialect merges with the rest of the dialect network. Last, but not least, at $w = 0.75$, the southern dialect chain AM–AG–VLR emerges as an isolate dialect area eventually connecting to the bulk of the dialect network only at $w = 0.50$ through a double bind with peripheral TP to the west (only AM and AG, while VLR remains disconnected). While the previous configurations of graphs pointed at transactional networks (through morphosyntax) and settlement networks (through phonology), the graph sets based on lexical variation can be viewed as the product of congregative networks—a subtle compromise between transactional (i.e. contact) networks and historical (settlement) networks. Congregative networks make up compartmentalized feature pools, providing robust cells or variation fields, from where innovative trends in grammar and speech may spill out and exert skewness.

3.7.5 Analysis of the Tseltal database: Comparison

By comparing the alternative standpoints of asymmetric or complementary taxonomies of the Western Tseltal dialect network considered in our study, we can observe certain relevant points.

First, when analyzed separately, grammar, phonology, and lexicon reveal different pictures of the inner structure of relationships (or interactions) within the dialect network. While morphosyntax (discourse) points at a confederation of varieties taking place in the center and the north of the area, phonology (pronunciation) points at a tighter and more intricate alliance of the center and the south. On the other hand, lexicon (words) points at a more even, threefold, and densely connected distribution of dialect areas.

Second, the web of interactions and relationships unravels fine-grained properties of the communal aggregates, which emerge from different thresholds, according to various phenomenologies; for instance, it sheds light on the role of varieties such as Oxchuk (OX) or TP (Tenejapa), Tenango (TG) or Guaquitepec (GU) as intermediary nodes.

Third, properties of the language components show that these networks differ in their typology from the standpoint of social history and geography (external factors). Grammar and discourse networks may well be dependent on transactional webs of interactions between speakers and speech communities; however, phonological networks may depend more on the consolidation of speech style and prestige within historical communities, clinging more or less to settlement patterns.

In this respect, the contributions of historians such as Reyes García (1962) and de Vos (1990, 1992, 2010) shed light on the intricate web of factors and events that may account for the diversity of the Western Tseltal dialect network and many others over Central and Southern America. Not only did the demographic depletion of the native population all over America after the Conquest (Benassar 1980) have dramatic effects on language and culture, the scattered settlements of the pre-Columbian period in scarcely inhabited ceremonial centers were also modified through *reducciones* (concentration settlements) in order to politically control the natives and contract labor for the *encomiendas* and to serve the landlords of European origin. The dynamics of the political and territorial story of indigenous peoples under European rule in early colonial Mexico is a social analog of the dynamics of the cardiac systole (contraction) and diastole (expansion) cycles. While, under the Europeans, Indians were forced to cluster around the *fincas* (plantation farms) or in new urban centers, natives constantly tried to escape the agrarian slavery of *encomiendas* and *haciendas*, rushing back to their *milpas* (cornfields) and resuming their former scattered settlement patterns. Many intermediate solutions were tried throughout the history of colonial time, such as reaccommodation from one place to another or back to initial settings, after a demographic depletion caused by, for example, an epidemic. Moreover, in the eighteenth century, peasant (mostly indigenous) rebellions had severe demographic consequences, especially in the center of the Tseltal area, in Chiapas (Klein, 1970); the core of the rebellion happened to emerge in Cancuc (CA), a village linguistically tightly linked to Oxchuk (OX). The repression of the rebellion caused many casualties among the Tseltal population, especially among the elites, and many families had to move as refugees to other villages; political alliances between villages underwent drastic changes. For all these reasons, one should not consider the Western Tseltal dialect network studied here in continuity with the pre-Columbian times: it is rather a linguistic space, the complex output of many processes taking place against the backdrop of a continuous multivariable thread of demographic, territorial, and political changes. The tools of complexity theory may not be able to provide a Time Machine that can reach the linguistic communities far in the past. Yet, they can unravel many interesting aspects and trends of what happened in the last centuries; as long as links are not overinterpreted, they can tell a lot about how languages evolve in space and time.

The northern dialect chains that were seen as subgraphs point to communal aggregates (or population clusters, dialect communities, or grammatical and speech norms), which developed through *fincas* strongly linked with cattle breeding, around important economic centers such as YA, where coffee plantations were also on the rise at the end of the nineteenth century. The power of the church was strong in urban centers, such as CHI and BA. As suggested by the graphs, peripheral villages of the southern part of the north, like ST or GU were more rural transitional centers than leading nodes in the territorial web of local economy. The central dialect area benefited from the proximity of San Cristobal de las Casas. The Tseltal population atlast found a compromise between agriculture (the *milpa* traditional economy), trade, and transportation, especially in a urban center like Oxchuk. However, localities such as CHA or SP remained more isolated, thriving on a subsistence economy that revolved around the *milpa*. The case of the three erratic localities in the south (AM, AG, and VLR) can also be explained as a result of a combination of internal factors (the threshold graphs hinting at structural similarity) and external factors (social history, economy, and geography). These localities do not make up a web on their own as the patterns observed in the graphs suggest. On the one hand, their tropism differs from the central and northern dialect areas; on the other hand, their contact history with the Mestizo or Ladino demographic component of the Chiapas population also differs: AM and AG have undergone less acculturation, less language shift toward Spanish than VLR in the south, and have managed to reach some degree of autonomy and prosperity through the subtle balance of successful handicraft (pottery) and subsistence economy. The line–node chain AM and AG forms with TP in the central western corner of the dialect network, and PE in the northwest, as can be seen at threshold $w = 0.5$ in Figure 3.20 hints more at structural affinities as a kind of average peripheral type than at genetic affinities proper.

Chapter 4

Historical Glottometry

4.1 Background

The family tree model (*Stammbaumtheorie*) is a major approach to the interpretation and visualization of the (historical) relationships among languages. It provides a simple and clear picture of the emergence of a language family, starting from the single original community speaking the protolanguage, then gradually splitting up into more and more subcommunities, finally turning into the observed group of languages.

One of the main limitations of the tree model of language evolution is the underlying assumption that the protolanguage develops *independently* in each branched subcommunity. Such an idealized situation rarely occurs; usually, innovations are born in one community and spread to other adjacent communities. The diffusion of innovations across a dialect continuum can result in a patchwork of intersecting innovations, a scenario that does not fit into the simple pattern of a tree model and a cladistic approach. To overcome the difficulties of the tree model, the *wave theory* was proposed and developed in the framework of historical linguistics (see de Saussure [1915] and Bloomfield [1926]). The wave theory allows the description of languages as entities continuously evolving while interacting with each other. To distinguish between these very different situations described by a tree model on one hand and wave theory on the other hand, the term *language family* was proposed in a sense similar to the genetic family, that is, as isolated subgroups (Ross, 1988). The term *linkage* was also introduced to represent a language inside an interactive language network (Schmidt, 1872), in which, in principle, all languages interact with each other.[1]

---

[1] Notice that a similar term 'linkage' is used in genetics, but with the different meaning of that *property* of single genes that is passed together to a new cell DNA.

4.2 Wave Theory

In this section, we focus on *historical glottometry*, which was inspired by the wave theory and introduced in the works of Siva Kalyan and Alexandre François on the languages of Vanuatu, an island nation in the Australian continent (François, 2012). The basic idea of the wave model is the evolution and spreading of innovations, which François defined as *changes* shared by all the languages of modern speakers (François, 2014). The presence or absence of innovations creates geographical borders known as *isoglosses* which define the areas corresponding to languages, which, in the spirit of wave theory, usually overlap with each other.

The starting point of historical glottometry is the dialect-based comparison of the innovations, namely the distinction, for a given pair of languages A and B, between:
- *Exclusively shared innovations*, meaning those owned by both language A and B, with respect to another language C.
- *Conflicting innovations*, that is, those innovations which are different.

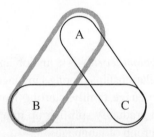

Figure 4.1 Historical glottometry: An example of three intersecting subgroups A, B, and C. (From Rannap [2017].)

From the number p of shared innovations and the number q of conflicting innovations, one computes the relative strength K of the grouping between a language A and a language B, named *cohesiveness*, as

$$K_{AB} = \frac{p}{p+q} . \tag{4.1}$$

A cohesiveness value $K = 1$ means that all innovations are common to A and B, while $K = 0$ means that there is no common innovation among A and B. As an example, let us consider a dialect family consisting of three languages A, B, and C, in which isoglosses define intersecting groupings, with 12 innovations exclusively shared between A and B, 4 between A and C, and 2 between B and C, as schematized in Figure 4.1. Then, the cohesiveness of the subgroup AB is

$$k_{AB} = \frac{12}{12+4+2} = \frac{2}{3} = 0.67 . \tag{4.2}$$

meaning that 67% of the innovations attest the cohesion of the subgroup, while 33% contradict it.

Additionally, historical glottometry defines the *subgroupiness* s_G of a subgroup G. If a subgroup G has cohesiveness K_G and a number ϵ of exclusively shared innovations, then its subgroupiness is calculated as

$$s_G = \epsilon K_G .\tag{4.3}$$

Let us consider the example shown in Figure 4.1, in which the subgroup AB has 12 exclusive innovations and a cohesiveness $k_{AB} = \frac{2}{3}$ (Eq. [4.2]). The subgroupiness of AB is

$$s_{AB} = \varepsilon \cdot k_{AB} = 12 \cdot \frac{2}{3} = 8 .$$

This gives us the strength of the subgroup AB. The subgroupiness of AC and BC can be calculated similarly,

$$s_{AC} = \varepsilon \cdot k_{AC} = 4 \cdot \frac{4}{12+4+2} = \frac{16}{18} = \frac{8}{9} \approx 0.88,$$
$$s_{BC} = \varepsilon \cdot k_{BC} = 2 \cdot \frac{2}{12+4+2} = \frac{4}{18} = \frac{2}{9} \approx 0.22.$$

To visualize the subgroups, we can now draw lines around the subgroups that are directly proportional to the subgroupiness of every given subgroup. We call these diagrams historical glottometric diagrams.

Two more examples are given in Figure 4.2. (the corresponding parameters are provided in the caption).

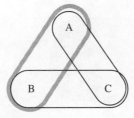 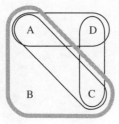

Figure 4.2 Historical glottometry: Two examples of three intersecting subgroups A, B, and C, with the following subgroupinesses: $s_{AB} = 8, s_{AC} = 0.88, s_{BC} = 0.22$, (left figure); $s_{ABC} = 8, s_{AC} = 0.88, s_{AD} = 0.68, s_{DC} = 0.28$ (right figure). (From Rannap [2017].)

Historical glottometry differs in important aspects from distance-based methods. First, a 'distance' is supposed to quantify the difference itself between two entities, such as two words, in different languages. Cohesiveness does not measure distances; it starts from an assigned set of innovations and compares the exclusive innovations with the shared ones. Furthermore, the criteria of historical glottometry employ quantities, which already contain information about the whole group; this is in variance with those of distance-based methods, which make direct pair comparisons, that is, they compare two languages at a time, in the hypothesis that further comparison between all the distances thus computed will shed light on the subgroups of the system.

4.3 Case Study of Numic

In this section, we consider the case study of Numic languages and report the results of a study (Rannap, 2017), based on various methods, including historical glottometry, aimed at revealing the structure of the Numic language network and its compatibility with the hypothesis of the Numic spread hypothesis.

4.3.1 The Numic spread hypothesis

The Numic languages belong to the Uto-Aztecan language family that is widely distributed in the western United States and Mexico (Shaul and Ortman, 2014). They are the northernmost subdivision of it, being situated in northwestern United States; the languages are concentrated in the Great Basin, which is a watershed in a set of smaller basins between generally high tablelands, bounded by the subrange of the Rocky Mountains in the east and the Sierra Nevada mountain range in the west (Grayson, 2011). The geographical distribution of Numic languages is depicted in Figure 4.3; it is based on the Native Land website.[2] Like the other Uto-Aztecan languages, the Numic languages are thought to have originated from a single protolanguage. The speakers of the Proto-Numic were by nature highland people, originating from the southern

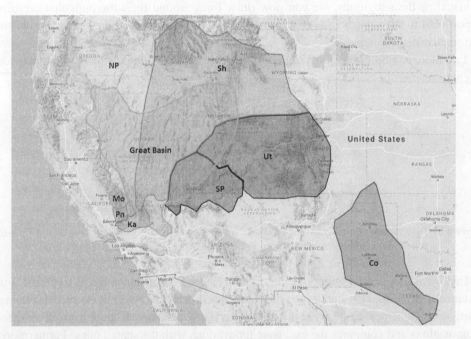

Figure 4.3 Map of the Numic area, showing also the Great Basin. The Numic languages considered are as follows: Co = Comanche, Pn = Panamint, Sh = Shoshoni, Ka = Kawaiisu, SP = Southern Paiute, Ut = Ute, Mo = Mono, NP = Northern Paiute. (From Rannap [2017].) (Map data © 2019 Google, INEGI.)

---

[2] http://native-land.ca.

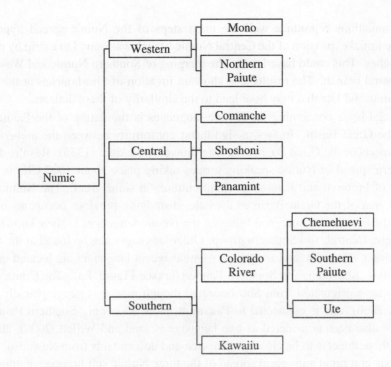

Figure 4.4 Tree of Numic languages illustrating the composition of the Western, Central, and Southern groups. (From Rannap [2017].)

part of the Sierra Nevada mountain range and Death Valley, also called the Numic homeland, in/near the southwest corner of the Great Basin. The Numic languages make up a dialect system consisting of seven main languages that are divided into three subgroups based on their geography: Western Numic languages, Central Numic languages, and Southern Numic languages (Grayson, 2011) (see the map in Figure 4.3 and the language tree in Figure 4.4). Notice that the Colorado River language is divided into three dialects, but in this section, we consider only Southern Paiute and Ute.

Diffusion of language can occur through the movement of population (Bettinger and Baumhoff, 1982; Madsen, 1975); a similar mechanism seems to have influenced the development of the Numic group structure. Lamb (1958) proposed the Numic spread hypothesis that the Numic languages followed the diffusion of the speakers from the original Numic homeland across the Great Basin, around 1000 CE, resulting in a fan-like distribution of the speakers into seemingly three groups. Lamb argued that the division of the groups appeared while the Numic speakers were still inhabiting their homeland (Figure 4.3), whence a disturbance triggered the rapid northeastern spread. The rapid spread of Numic speakers is an anomaly in the evolution of Numic dialects, because there seems to be no starting cause (Bettinger and Baumhoff, 1982). This interesting case has enthused researchers of different fields, from linguistics to computer sciences, to search for the missing factor. In fact, the Numic diffusion process has also been studied through computer simulations to reconstruct the competing populations in the Great Basin area (Young and Bettinger, 1992). The results from the simulations reproduce well the main steps of the Numic spread hypothesis, for example, the initial expansion of the Central Numic branch was bound to a strip by neighboring dialect branches. This could have lead to the merging of Southern Numic and Western Numic with the Central branch. The results also show an invasion of Shoshoneans at the expense of Northern Paiute and Ute that may have lead to the similarity of these dialects.

One crucial issue concerning the Numic languages is the dating of the Paiute–Shoshoni pottery in the Great Basin. In fact, a significant conformity between the archeological and linguistic aspects of the Great Basin area was shown by Madsen (1975). Results demonstrate a northeastern spread of Numic speaking groups taking place about 1000 CE, as well as the coexistence of Fremont and Paiute–Shoshoni cultures in some areas. The Numic expansion affected the area of the Great Basin as a whole, dislodging previous occupants of the Great Basin, like the Fremont culture, and bringing the Numic languages to their known groupings of the Western, Central, and Southern group. Older languages can be found at the roots of the branches: Mono, Panamint, and Kawaiisu; whereas recent languages are located in the stems: Northern Paiute, Shoshoni, and Southern Paiute/Ute (see Figure 4.3). The Comanche dialect is known to have originated from Shoshoneans, though now it is geographically far (Casad and Willett, 2000), and is connected to Panamint, as is Shoshoni. Southern Paiute and Ute dialects have also been considered as one language (Casad and Willett, 2000), allowing one to consider these dialects to be closely connected and descendants from Kawaiisu. In general, the hypothesis of a rapid and recent spread of the three Numic sub-branches comprehensively matches the wedge-shaped distribution of the Numic people (Bettinger and Baumhoff, 1982; Hopkins, 1965; Young and Bettinger, 1992). We can thus expect the dialects in the geographical subgroups (Western, Central and Southern Numic) to be more similar in both phonological and morphological sense than those of neighboring branches.

4.3.2 The Numic database

The first part of the analysis of the Numic languages presented here concerns the preparation of the database to be analyzed. Starting with the original compilations of Davis (1966) and Iannucci (1972), merged into a new database by Dalmasso et al. (2012), some preliminary editing work was carried out with a twofold goal.
- To homogenize the transcriptions between the two sources since different phonological theories were used (i.e., some sounds were represented differently between the entries)
- To remove the affixes of the entries and reveal the roots of the words.

The modified database thus obtained consisted of 292 cognates of the protolanguage, that is, words with a similar etymological origin, and their respective counterparts (entries) in the eight largest Numic dialects, all the data being in IPA transcription (Rannap, 2017). The cognates were constructed based on a hypothesis of what the words in the dialects may have been—thus it is not a full Proto-language once spoken by the Pre-Numic people—using phonological and grammatical rules known to be valid in other languages. Unfortunately, not all languages are fully represented in the database for every cognate. Table 4.1, which gives the number of cognates in each language, provides an overview of the size of the set of words used for comparing dialects and calculating Levenshtein distances.

Table 4.1 Number of cognates in the languages of the Numic database, 292 cognates in total (abbreviations as in Figure 4.3).

Dialects	Co	Ka	Mo	NP	Pn	Sh	SP	Ut
Co	206							
Ka	45	56						
Mo	142	55	189					
NP	161	53	153	198				
Pn	67	27	59	59	86			
Sh	188	48	157	174	70	231		
SP	164	54	159	159	71	178	220	
Ut	64	32	62	62	34	63	66	72

4.3.3 Results from historical glottometry

The basis of historical glottometry is the dialect-based comparison of innovations, which in the case of the Numic languages are phonological innovations. One can make these innovations apparent using the Levenshtein distances. By accounting for the changes made to every word pair, one can attain specific changes made to words in all eight Numic dialects. To reduce biased results, as Levenshtein distance in accounting for changes was ambiguous in some cases, the changes which occurred less than five times were discarded; only the changes occurring more than five times, referred to as *regular innovations*, were taken into account. Rannap (2017) used the Levenshtein distance method to identify 32 phonological innovations. From those 32 innovations, only 30 subgroups shared any exclusive innovations. This would give biased results when calculating the subgroupiness. To alleviate this problem, one needs to introduce irregular innovations. In contrast to regular innovations that occur in every case of a specific context, irregular innovations are found only in one word or a small group, providing, therefore, a more empiric result. Examples of irregular innovations can be seen in Table 4.2, where in each row, languages of the same color represent the specific irregular innovation.

Table 4.2 Examples of irregular innovations.

	Co	Ka	Mo	NP	Pn	Sh	SP	Ut
piya	pia?	piya	piya	pia	piya	pia	pia	pia
pəhta	pəda	pəda	pəta	pəta		pəda	pəda	
suwah	sua		suwa		suwa	suwa	sua	səa
səhə	səhə	səə	səhə	sə		soho	səə	
tahma	tahmani		tawano	tamanu	tahwani	tahmani	tamana	tamana
taman	tama	tawa	tawa	tama	tama	taŋʷa	taŋʷa	
tape	tabe		tabe	taba		tabe	taba	
...	...	...	...	...	...	...	...	...

While accounting for the irregular innovations, 239 innovations were received in addition to the 32 regular innovations calculated from the Levenshtein distance. For the final lexical innovations, the regular and irregular innovations were put together. On the basis of the 271 innovations, Table 4.3 was compiled with the occurrences of binary values: the value 1 accounted for the presence of an innovation in a language, and the value 0 for the absence of an innovation in a language. From the table of occurrences, one can calculate the cohesiveness and exclusively shared innovations for all dialect subgroups; from these values, the subgroupiness ς for all sets of the Numic dialects can be derived. The ten strongest subgroups of the Numic dialects are shown in Table 4.4.

Table 4.3 Examples of regular and irregular innovations.

	Co	Ka	Mo	NP	Pn	Sh	SP	Ut
aa ↔ a	1	1	1	1	0	0	1	1
ə ↔ a	1	1	0	0	0	0	1	0
k ↔ q	0	0	1	0	0	0	1	0
hk ↔ k	0	1	0	1	1	1	0	0
hm ↔ m	1	0	1	1	1	1	1	1
irr1	1	0	0	0	1	0	0	1
irr2	1	0	1	0	1	0	0	0
...	...	...	...	...	...	...	...	...

Table 4.4 Ten strongest subgroups of Numic dialects.

Subgroup	ε	k	subgroupiness (ς)
Co–Sh	27	0.458	12.363
Mo–NP	22	0.410	9.011
SP–Ut	13	0.323	4.194
NP–Sh	12	0.340	4.075
Co–NP	9	0.361	3.249
Co–NP–Sh	8	0.240	1.923
Sh–SP	6	0.299	1.791
Co–Mo	7	0.252	1.767
Co–Mo–NP–Sh	9	0.160	1.440
Mo–SP	6	0.235	1.408

The glottometric diagram for the Numic dialect system is presented in Figure 4.5. The width and color strength of the isoglosses plotted are proportional to the strength of the subgroups: the thicker the line width and darker the color, the stronger is the subgroup. The diagram shows four strong dialect pairs: [Co–Sh], [NP–Mo], [SP–Ut] as well as [Sh–NP] subgroup.

The strong [Sh–Co] and [Ut–SP] links are consistent with the fact that Shoshoni and Comanche, as well as Ute and Southern Paiute, have also been considered as single dialect groups. The group [NP–Mo] is known to form the Western Numic languages group. While the similarity between Ute and Southern Paiute as well as between Northern Paiute and Mono seems natural due to their geographical vicinity, the similarity between Shoshoni

Historical Glottometry

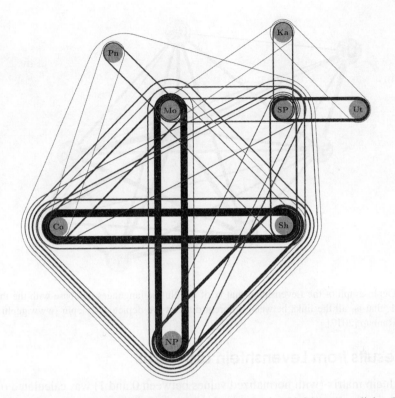

Figure 4.5 Glottometric diagram of the Numic languages, realized using the 24 strongest links. (From Rannap [2017].)

and Comanche might be unexpected when considering solely the geographical isolation of Comanche (see Figure 4.3). However, this is understandable because the Comanche and Shoshone people used to form a single group, from which the Comanche emerged as a distinct group only shortly before 1700, when they broke off from the Shoshone people [4]. Though at first sight the strong link in [Sh–NP] might seem unexpected, it is actually rather natural due to the long geographical continuity between the corresponding areas.

As a whole, we can see a system of four strongly connected dialects [Co–Sh–NP–Mo] and the aforementioned [SP–Ut] grouping with Kawaiisu and Panamint connected only by weaker isoglosses to other languages. The subgroup [Co–Mo–Sh–NP] also tends to link to the Southern Paiute. The subgroup [Co–Sh–SP] or the slightly weaker [Co–Sh–NP–SP] may proceed from the [Co–Sh] dialect group being weighed down between the Northern and Southern Paiute as speculated by Young and Bettinger (1992). The diagram also reveals that though Panamint is expected to have a strong connection with [Co–Sh], it is instead more strongly connected to Mono. However, this [Pn–Mo] connection is most probably due to the fact that these dialects are geographically neighbors forming the Western Numic dialects group. The Kawaiisu dialect, in addition to being linked to the Southern Paiute, is linked to the Mono dialect as was Panamint, which is supported in some way by the Levenshtein distance result where the Kawaiisu connected strongest to the Northern Paiute (in the same subgroup with Kawaiisu).

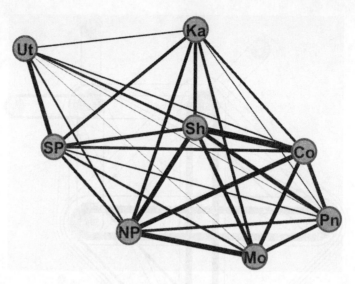

Figure 4.6 Gephi graph of the Levenshtein matrix of the Numic languages database with the threshold $T = 1$, that is, all the links between different dialects are depicted. (Gephi [www.gephi.org]) (From Rannap [2017].)

4.3.4 Results from Levenshtein distances

The Levenshtein matrix (with normalized values between 0 and 1) was calculated on the basis of the database described in this chapter and presented in Table 4.5 in the form of a heat map. The Levenshtein matrix is visualized as in Figure 4.6 in the form of a network.

Table 4.5 Matrix of weighted Levenshtein distances for eight Numic dialects.

Dialects	Co	Sh	Pn	NP	Mo	SP	Ka	Ut
Co	0	0.364	0.610	0.518	0.611	0.724	0.762	0.871
Sh	0.364	0	0.602	0.506	0.617	0.693	0.673	0.759
Pn	0.610	0.602	0	0.751	0.667	0.868	0.928	0.983
NP	0.518	0.506	0.751	0	0.470	0.725	0.666	0.820
Mo	0.611	0.617	0.667	0.470	0	0.783	0.723	1
SP	0.724	0.693	0.868	0.725	0.783	0	0.700	0.568
Ka	0.762	0.673	0.928	0.666	0.723	0.700	0	0.875
Ut	0.871	0.759	0.983	0.820	1	0.568	0.875	0

The thickness of the links increases as the corresponding Levenshtein distance decreases, so that one has an overview of the major links in the network. The visualizations are carried out using the Gephi software (gephi.org). The same Levenshtein matrix can also be visualized as in Figure 4.7 by introducing a (variable) threshold T, such that the link between a node i and a node j are shown only if the Levenshtein distance $L_{ij} < T$. The major connections between languages appear gradually, as T increases from $T = 0$ to $T = 1$.

Historical Glottometry

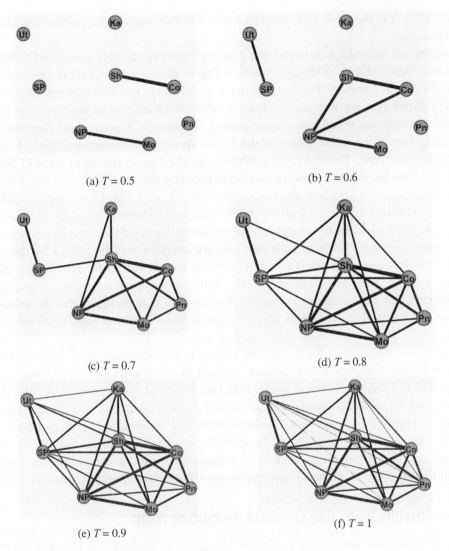

Figure 4.7 Gephi graphs of the Numic database with different thresholds T. (Gephi [www.gephi.org]) (From Rannap [2017].)

The strong links in [Sh–Co] and [NP–Mo] show that these language pairs are the most closely related ones in the phonological sense. There is also a strong link in [Ut–SP]. The second one arises from the link [Sh–NP] due to the already discussed fact that the Shoshone and Comanche people formed a single group in the past.

The link between Northern Paiute and Shoshoni/Comanche, which is considerably stronger than the link between Panamint and Shoshoni/Comanche, as well as the link between Mono and Shoshoni/Comanche, raises grave doubts about the thus far accepted division of the Numic languages into Central and Western Numic languages (see Figure 4.4). Indeed, our analysis suggest that the Central and Western Numic languages form one single group. A rather surprising result is the weak link between Mono and Panamint languages that are neighbors geographically.

Another unexpected result is that the Kawaiisu is more strongly connected to Northern Paiute and Shoshoni than to Ute and Southern Paiute although the dialect is believed to be a part of the [Ut–SP] branch. Furthermore, for $T = 0.7$, there is a link between Shoshoni and Southern Paiute (but not Kawaiisu). These results do not align with the former hypothesis of the Numic expansion and imply more contact between the Kawaiisu and Northern Paiute languages. Due to the recent nature of the Numic spread, this contact was likely before the expansion took place. Therefore, we speculate that the Numic spread to areas of Southern Paiute was from Shoshoni, not from Kawaiisu as formerly assumed.

We also notice that Shoshoni and Comanche occupy a central position in the network. This can be put into correspondence with the central location of Shoshoni (actually, it is due to the strong link, [Sh–Co], of the original group Shoshoni/Comanche) in the Great Basin with respect to the other languages. Such an emerging network structure may be caused by the Central Numic branch being compressed at the start of the Numic spread by neighboring dialects, as speculated by Young and Bettinger (1992). With Panamint having weaker connections to other dialects than Shoshoni and Comanche, it is possible to hypothesize that the interactions between [Sh–Co] and other dialects happened during the diffusion process of Numic and not before.

The relative isolation of Kawaiisu with respect to the other languages (even to Ute and Southern Paiute, with which it is usually grouped in the Southern Numic group) is to be noticed. The first links of Kawaiisu appear only for a threshold $T = 0.7$ (therefore, with a rather low strength), when all the languages have already become connected in a single–component network. The linked languages are Northern Paiute and Shoshoni, rather than the languages of the Western Numic branch (Ute and Southern Paiute), in which Kawaiisu is usually placed. This fact and other peculiarities, such as an atypical consonant inventory, may be due to the original Kawaiisu homeland being bordered by non-Numic Uto-Aztecan languages.

4.3.5 Results from the nearest neighbor map

The nearest neighbor network of the Numic languages is presented in Figure 4.8; for each node, only the link associated with the shortest Levenshtein distance of the node is represented. The position of the language nodes are arbitrarily chosen at the center of the respective area for the sake of visualization. Such a network is expected to represent the links between the most similar languages. In fact, one can see the expected groupings of Central and Western Numic branches, with the addition of the established [Ut–SP] chain. The only exception is the Kawaiisu dialect, which links closest to Northern Paiute. By embedding this network in the geography, one can account for the geographical influence of the territory on the spread of Numic dialects.

Historical Glottometry

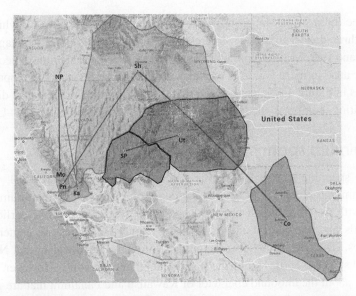

Figure 4.8 Nearest neighbor network of the Numic languages, in which each language is linked only to its closest neighbor. (From Rannap [2017].) (Map data © 2019 Google, INEGI.)

4.3.6 Results from the dendrogram

Among the various hierarchical clustering methods available, the dendrogram can be constructed for our database using Ward's minimum variance method; it was realized using the R Language for statistical computing (r-project.org). The dendrogram resulting from the Levenshtein distances of the Numic languages is shown in Figure 4.9.

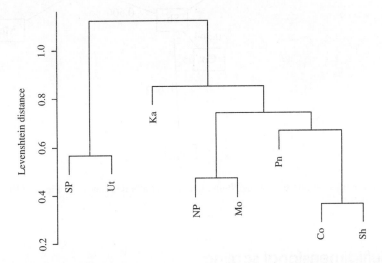

Figure 4.9 Dendrogram of the Numic languages (Co = Comanche, Pn = Panamint, Sh = Shoshoni, Ka = Kawaiisu, SP = Southern Paiute, Ut = Ute, Mo = Mono, NP = Northern Paiute). (From Rannap [2017].)

One can see three strong subgroups [Co–Sh], [NP–Mo], and [Ut–SP], which implies the coherence of three dialect branches. Interestingly, the Panamint dialect linkage is present in both [Co–Sh] and [NP–Mo] groups, which may mean that the Panamint dialect blended with northern Mono and Northern Paiute as well as with Comanche/Shoshoni. Comparing the Gephi network graphs in Figure 4.6, it can be seen that the dendrogram indicates the distant nature of Kawaiisu, locating it closer to the [NP–Mo–Pn–Co–Sh] chain than to the [SP–Ut] group.

4.3.7 Minimum spanning tree

The minimum spanning tree of the Numic languages, given in Figure 4.10, was built with Prim's algorithm and plotted with the R Language. In general, this minimum spanning tree is consistent with the information provided by the analyses made in the earlier subsections. The tree reveals how the Shoshoni and Northern Paiute languages are centered among the other languages. The tree also shows the connection between Kawaiisu and Northern Paiute (Levenshtein distance 0.666), as did the nearest neighbor network. However, there remains some unclarity about the unbound character of Kawaiisu and Panamint that are geographically neighbors.

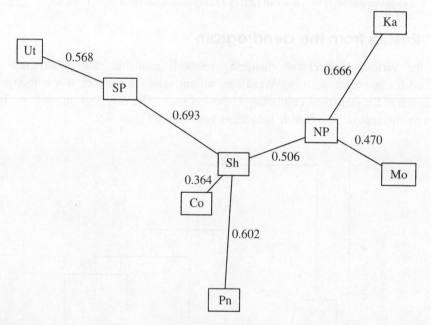

Figure 4.10 Minimum spanning tree of the Numic languages (abbreviations as in Figure 4.9). (From Rannap [2017].)

4.3.8 Multidimensional scaling

Multidimensional scaling of the Levenshtein distances of the eight Numic languages is shown in Figure 4.11. The Numic languages are font-coded according to the traditional divisions

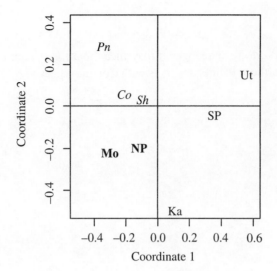

Figure 4.11 Result of the multidimensional scaling on a plane (abbreviations as in Figure 4.9). [From Rannap (2017).]

of Western (bold), Central (italic), and Southern (normal) languages. As can be seen in Figure 4.11, the results from the multidimensional scaling match, in general, the traditional divisions, with the exceptions of Panamint (which appears to be further from Shoshoni than Northern Paiute) and Kawaiisu (which is the most isolated language). This corresponds very well to the results implied by the minimum spanning tree (see Figure 4.10) as well as the other analyses presented earlier.

4.3.9 A short comparison

When comparing the results of the Levenshtein distance and historical glottometry, we see, besides the [Co–Sh], [SP–Ut], and [NP–Mo] subgroups, the confirmation of a strong dialect foursome, consisting of Comanche, Shoshoni, Northern Paiute, and Mono dialects. This grouping may be the cause of greater interactions between the four languages and is not entirely in contradiction to the Numic spread hypothesis. The other similar result between the two methods is the aforementioned Kawaiisu and Mono connections as well as the centered nature of the [Sh–Co] dialect group, which, as mentioned earlier, could be caused by the central Numic branch being compressed at the start of the Numic spread by neighboring dialects (Young and Bettinger, 1992).

In addition to the confirmed results, the two methods give different results for the Panamint and Kawaiisu dialects. Using historical glottometry, we can see that Kawaiisu is connected to Southern Paiute as expected and not to [Co–Sh] as seen from the Levenshtein distance method (see Figure 4.6). For the Panamint dialect, it is the reverse case, with the Levenshtein distance method producing results in accordance with the Numic spread hypothesis, but not strongly supporting the connection to the Mono dialect, seen in the glottometric diagram (Figure 4.5).

PART TWO
LANGUAGE DYNAMICS

Chapter 5

Introduction to Language Dynamics

In this chapter, some mathematical models of language dynamics (Wichmann, 2008a,b; Schulze et al., 2008; Vogt, 2009; Castellano et al., 2009; Solé et al., 2010) are discussed and illustrated with some applications. Many language dynamics models have been developed to describe, for example, the evolution of languages, the competition processes between different linguistic features (considered as fixed entities), or the cognitive dimension of language. Due to limited space, it is not possible to provide here an exhaustive discussion of language dynamics models and many important pieces of the complete picture of the field are missing. In the following sections, we will discuss a selection of examples with the goal of providing at least a general idea of the field and why it represents a framework for further possible developments.

The ultimate goal of mathematical language dynamics models is to provide a quantitative description of language change, that is, of the combined dynamics of evolution, competition, and spreading processes of languages in space and time and of the consequent diversity and correlations of the linguistic landscapes, discussed in the first part of the book. For convenience, models are classified into different types. Fully evolutionary models, in which languages themselves undergo changes, while possibly competing with each other, are discussed in Chapter 6. Competition and natural selection models, which consider the processes taking place on a shorter time scale, and on which language features can be considered fixed, are discussed in Chapter 7; they can in turn be classified into models studying the time scale of language shifts and those focusing on the shorter time scale of language use. We explore the important and interesting cognitive dimension of language only partially, by reviewing a few semiotic dynamics models, which are discussed in Section 6.1.

5.1 Motivations behind Language Dynamics Modeling

There are different motivations that drive the study of language dynamics, ranging from scientific to social ones. These diverse motivations reveal the interdisciplinary nature and wide applicability range of language dynamics.

- Language represents one of the most complex known phenomena. This fuels the interest of scientists who are keen to solve life's every problem, be it complex or simple.

- An interest in the *past history of languages* is related to the fact that reconstructing the history of languages provides crucial information that helps to understand the general history of, for example, the cultural landscape and movement of populations. By using mathematical models and performing complex data analysis of linguistic databases, language dynamics can provide the opportunity to reconstruct historical initial conditions and time evolution processes that often cannot be investigated otherwise due to the lack of information, such as, documents or archeological findings.
- There is a growing interest in the *future of languages* from a twofold social perspective:
    (i) Preserving endangered languages, not only to prevent the loss of the languages themselves, but because they are linked to corresponding endangered cultures; these cultures represent sources of high cultural diversity, which are now recognized as very valuable cultural heritage that are to be passed on to future generations (Sutherland, W. J., 2003, Nature 423, 276; Krauss, M., 1992, Language 68, 4). The consequent relevance of understanding how such a cultural extinction process may take place is well summarized by Mende and Wermke (2006):
    Understanding cultural evolution is one of the most challenging and indispensable scientific tasks for the survival of humankind on our planet. This task demands, besides an adoption of theories and models from biological evolution, theories for culture-specific processes as well. Language evolution and language acquisition offer interesting objects of study in this respect.
    (ii) Global and vehicular languages that will have crucial roles as *linguae francae*, allowing people to communicate effectively with each other and enjoy the benefits of accessing updated and exhaustive sources of information.

It should be noted that the latter points are not necessarily in contradiction with each other: in a society of bilinguals or multilinguals, one can find both one lingua franca or a few linguae francae, on one hand, as well as some local languages, on the other hand, with a degree of linguistic diversity that may be the optimal adaptation response to the needs of a fast and effective communication system used by individuals across all the levels of society.

5.2 Numerical Calculations and Experiments

There is a strong computational component underlying the study of language dynamics and it is useful to start by considering such a computational side from a general perspective. In fact, numerical procedures, that is, all the computational methods used to carry out calculations with the help of different tools, such as computers, mechanical calculators, the simple slide rule, or even by hand, distinguished from their analytical counterpart, that is, calculations based on mathematical laws and theorems, have always been a relevant part in any scientific discipline.

In the past—in fact until not so long ago—'computers' were people hired for doing the calculations, for example, for evaluating complicated integrals. Now, by 'computers' one only means electronic computers, which allow one to make remarkably bigger and more complex calculations. At some point, certain researchers had the idea that the new possibilities offered by computers for doing numerical computations could be used not only to evaluate the results of complicated formulas but also to solve the equations of the laws describing the behavior of a system, in order to check its evolution in space and time, that is, to make a *numerical experiment*. This way of using computers has opened a new approach to study natural phenomena. The first numerical experiment was the simulation of the *Fermi–Pasta–Ulam–Tsingou problem*, for studying the relaxation of a chain of nonlinear oscillators (Fermi et al., 1955). The interesting story of the unexpected results found and the merits of that work made it a cornerstone of various scientific fields such as nonlinear dynamics, statistical mechanics, and, of course, computational physics (Dauxois, 2008).

Eventually, numerical simulations and numerical experiments have become recognized as an actual alternative theoretical approach, complementary to the traditional one based on analytical calculations. In many cases, there is a sort of symbiosis between the two approaches, making it difficult to draw a line between them: numerical simulations are used to check the consistencies of theoretical frameworks and effective numerical methods are based on previous analytical results. This has happened not only in physics and hard sciences, but in many other disciplines, where numerical simulations are used extensively and in a systematic way to study a wide range of problems.

Numerical experiments occupy a peculiar conceptual position, located somewhere between analytical calculations and real experiments. This is because, as theoretical tools, they can be used in principle to reproduce the actual behavior of a real system if the underlying dynamical laws are known. This fact is strategical for studying different problems that are too complex for a simple analytical treatment, from validating or exploring the consequences of a theoretical model to the interpretation of the results of a real experiment. Moreover, in many cases, it is not about the complexity of calculations but about the mere impossibility to make a real experiment that replicates the history of the system under study, for example, cosmology in the hard sciences or the study of cultural and linguistic change in the social sciences. Often, it is not even possible to proceed through an approximate analytical study. In all these cases, resorting to numerical simulations can be a key strategy.

It is therefore natural to use numerical experiments to study linguistic systems. The general procedure used to construct and study models of language dynamics is the same used in other scientific disciplines. For example, a problem can be first approached along all lines which allow some understanding and provide some suggestions for constructing a simple model. Then the model goes through a series of gradual, successive validations and extensions toward more refined and general versions. Keeping models minimal in nature and in agreement with observations at each step of the investigation is a main underlying criterion of the procedure. This may result in oversimplified models describing data only within a limited parameter range. However, this is not to be considered as something negative in general, because simple models allow us to understand better some specific features of the underlying processes and deduce the right direction to take for further generalizations.

These considerations lead us to a characteristic of numerical experiments that is as peculiar as it is interesting: the implicit simplifications present in any model imply that 'all models are wrong' (Box, 1976), at least in principle; hence, numerical experiments usually explore the behavior of systems which do not (and sometimes cannot) exist. While at first sight this may seem paradoxical from the viewpoint of a scientific approach, it is this very circumstance that allows researchers to obtain reliable and consistent scientific results. In fact, every scientific prediction—even the most precise and successful one—is always statistical in nature, in that results are stated with a corresponding error, expressing uncertainty on the actual result. In the outcome of a real experiment, there is at best a most probable result, which is sometimes expressed as a mean value associated with an uncertainty. In a numerical experiment, the situation is similar and one needs to find the parameter set that leads to results as close as possible to those of the actual system, evaluating the corresponding uncertainties. A general statistical framework, which is suited for such quantitative comparisons, is that provided by Bayes statistics (Harney, 2003).

In some cases, it may be useful to study models designed to correspond to a fictitious world, even if the corresponding systems are, by definition, inaccessible to any real experiment. For example, one can be interested in studying a hypothetical world where electrons are heavier than the real ones to see how such a world would look like; this may provide some interesting information about the relevance of electrons having a specific value of the mass. One can also study human dispersal during the neolithic migration wave in different climatic conditions or with different underlying physical geographies to reveal the role of the various ecological factors on the dispersal process.

In the following sections, various mathematical models of language dynamics are discussed (not necessarily reflecting a historical order), with the goal of gaining an understanding of the main elements and features of a numerical model. This may be useful not only to construct a model and make some numerical simulations, but also to look critically at various models, to check their predictive power and limits, and possibly infer what changes are needed to provide a better description of the system under study.

5.2.1 Constructing a model of language dynamics

Many mathematical models of language dynamics resemble other models developed in different fields. This is due to the fact that they were inspired, for example, by physics models of statistical systems of interacting particles, ecological models of species competition, such as the Lotka–Volterra model, agent-based models in economics, or social sciences models, such as the geographical individual-based model of urban segregation of Schelling (1978), the voter model (Clifford and Sudbury, 1973; Liggett, 1999; Castellano et al., 2009), or the culture spreading model of Axelrod (1997).

The most relevant characteristic of a model is the degree of detail, or *coarse-graining* level, at which the model describes a system. Three general categories useful to define the *coarse-graining* level of a model are as follows.

(i) **Individual-based Models.** Starting from what is probably the most general framework for carrying out simulations of a complex system, individual-based models provide an exhaustive description of a complex system composed of a finite number N of units interacting with each other, in terms of the state of each unit in time. In language dynamics models, the units can represent the speakers, the linguistic groups/dialects of a population/language, etc. For the sake of clarity, here, we will refer to a complex systems model representing a speaking community, in which the units are interpreted as the individuals, that is, the speakers. Different terminologies are used in other fields to indicate concepts that overlap totally or partially with 'units' or 'individuals', such as the 'agents' of computer science and economy, the 'actors' of artificial intelligence, the 'bugs' of some ecology models or the 'Brownian agents' (Schweitzer, 2003). Since the features of a single individual can be specified arbitrarily, individual-based models are an ideal framework for the description of complex phenomena, such as size-dependent effects, fluctuations-induced effects (e.g., associated with the births and deaths processes of individuals), diversity-induced effects, as well as the effects due to the network topology underlying the interactions between individuals.

(ii) **Macroscopic Models.** By taking the limit $N \to \infty$, one can obtain the continuous limit of macroscopic models, in which the description is in terms of macroscopic variables. A typical example is an individual-based model of a population of N individuals speaking either language A or B, which in the limit $N \to \infty$ is described only by the time-dependent population sizes $N_A(t)$ and $N_B(t)$. An ecological example of transition from a micro- to the macroscopic model is the transition from an individual-based model of species competition to the corresponding Lotka–Volterra equation, which describes species in terms of their population sizes. There is no noise in these models. They are also called *mean field models*.

(iii) **Mesoscopic Models.** It is also possible to use a mesoscopic level of description, intermediate between micro- and macroscopic levels, in which macroscopic variables (such as the population sizes in the example described earlier) are introduced considering the case of large but finite populations $N \gg 1$, without taking the limit $N \to \infty$. In a mesoscopic description, the system is described by macroscopic variables (like in macroscopic models), but there are also noise terms that promote macroscopic variables to stochastic processes. If fluctuations (the stochastic terms) are neglected, one recovers the macroscopic deterministic description.

In the following sections, we consider mainly individual-based and macroscopic models.

Besides the level of coarse-graining, the form of the model will depend on the processes to be described. Here is a short list of processes that can enter a language dynamics model.

- Changes of state of an individual induced by other individuals; they can be, for example, a language shift, the adoption of a synonym or a linguistic feature from neighbors, the transformation from monolingual to bilingual, etc. This category of changes is described in language competition and selection processes. In *semiotic dynamics models*, a change of state may represent a change in the way meanings are associated with words.
- A change of state only due to the individual in question, for example, the spontaneous appearance of a (new) linguistic feature, an innovation, analogous to genetic mutations appearing in genomes.
- Demographic processes, related to birth and death processes (in turn, possibly related to competition for resources), immigration, and emigration.
- Dispersal, for example, the natural diffusion process of a population across a geographical landscape, which can also be related to immigration and emigration processes or other population movements induced by economic reasons.

It should be noted that the various processes listed here are in general constructs useful to define the conceptual and mathematical structure of the model system. In general, these processes do no act independently; it is their interplay that determines the system dynamics. For example, which linguistic innovations are made more frequently may depend on underlying cognitive processes; the initial and boundary conditions of the ecological environment may be crucial factors for the outcome of a language competition; the social network characterizing a linguistic group strongly influences the spread of a linguistic innovation; and the underlying population dynamics, which is a relevant factor in determining the outcome of a language competition process, may enhance (or start) large-scale dispersal processes.

5.3 A Minimum Dictionary

The models of language dynamics discussed so far, in particular, the language competition models associated with the presence of two competing languages and of a bilingual group, shows a rich variety of possible final scenarios and competition mechanisms, despite being, primarily, basic mathematical models based on simplifying (if not oversimplifying) assumptions. In order to proceed toward a more in-depth study of real situations, to deal with the actual complexity met in the study of language evolution and competition, and characterize properly the models and processes they define, it is useful to have a minimum linguistic vocabulary.

Natural Language. A language that has evolved naturally in humans through use and repetition, without conscious planning or premeditation (in contrast to constructed and formal languages such as those used to write computer codes or in logic).

Dialect Continuum. A set of languages that cannot be enumerated and divided into a group of definite dialects: their mutual linguistic distance decreases while their mutual intelligibility increases when their geographical distance decreases, so that one can go from one language to a different one by moving across a geographical region.

Lingua Franca. Lingua Franca is now understood to be a general language used to communicate between groups of people with different native languages for commercial, cultural, religious, or diplomatic reasons. Originally, *Lingua Franca* was the language spoken in the Mediterranean ports; it contained elements from different languages, such as Italian, French, Spanish, Greek, and Arabic.

Language. Any natural language may basically be defined as the association of a lexicon (i.e., a stock of words and categories such as entities, states, and processes (Jackendoff, 1997, 1990, 1985) and a grammar (functions and rules or constraints for subcategorization or association); in other words, a language is made of sets of items and combinatorics. Further properties to take into account resort to linearity, iterativity and embedding (recursivity), hierarchization, and specification of values or functions (Mufwene, 2001, 2013a,b,c), all amounting to a form of highly articulated technology resorting to a specific form of semiopoesis (that can be defined as the self-organization of semiotic patterns within cohesive and articulated systems of symbols), more complex than in any other tool for communication within other species on earth. Chomsky (1981, 1986, 1990) summed up these properties through a government and binding model, although the powerfulness of this two-word synthesis in the fabric of natural languages has been strongly undervalued. In Chomsky's terms, government accounted for the combinatorics of subcategorization—marking and dependency—while binding resorted to wider-range constituency, recursivity, and co- or cross-reference. These properties are modular, that is, distributed from the level of phonological form up to the morphosyntax level (Scheer, 2004, 2011, 2012). At a higher level, resorting to language proper, the logic form of relations between entities and states or processes and their topological constraints made up the language faculty in Chomsky's model. This view has often been misunderstood and considered as mentalist, although Noam Chomsky never denied that nothing more than the language faculty, as a cognitive device or predisposition in the human mind, is innate; it needs intense activation in the social context and through the use of natural languages to develop and reach a stable state. Chomsky's theory was not concerned with the conditions of this sociocultural exposure to specific languages, or with the intricacy of its complexification in multifold repertoires. This latter domain instead is explored by sociolinguistics (Mesthrie, 2011) and the field of languages in contact (Hickey, 2010).

Creole. Naturalized former pidgins emerging from language replacement for communal aggregates facing transplantation, especially in the tropical plantation economy since the seventeenth century are referred to as creoles. On the one hand, the former local or community languages can hardly be used efficiently anymore due to intricate miscegenation; on the other hand, the new language used in the new setting undergoes a thorough reanalysis. The emerging patterns can be described as follows:

  i) extended phonological primarization (neutralization of markedness);
  ii) massive lexical importation from the so-called parent language (French, Portuguese, English, Dutch, Spanish);
  iii) generalized grammaticalization, implying intricate patterns of structural reanalysis.

The question of whether creoles emerge from a substrate (African languages or even a Romance lingua franca from Portuguese in its early genesis; this theory is now considered as an unsustainable hypothesis) or only depends on a readjustment of the superstrate (the parent language) has been debated for long. Undoubtedly, creoles are the result of complex interactions between four kinds of networks of parameters: phylogenetic, universal grammar, typological, and areal variables. There may be few parallels of creole genesis in biology, as creole genesis implies models of self-organizing communal aggregates making up totally new systems out of transplantation in an alien setting (the parent language), on erasing all older lexical and grammatical patterns used formerly. This resilient situation ends up in a synthesis between cognitive constraints (Universal Grammar) and new patterns of naturalization (areal and social accommodation of communal aggregates, intergenerational transmission, and further expansion and complexification of the lexicon and grammar).

Diglossia. The term 'diglossia', formed with the Greek prefix *di-* ('two') and the root *glôtta/glôssa* ('tongue')—it can be compared with its Latin counterpart *bi-* + *lingualis* for 'bilingual'—was initially coined by Ferguson (1959), although the concept of diglossia had already a well-established tradition among hellenists before, with a meaning more or less equivalent to bilingualism. Diglossia is actually a kind of subordinated bilingualism of two languages X and Y, X having a lower status than Y. In diglossia, the s (status) parameter characterizing some language dynamics models, such as those presented in later sections, plays a crucial role, as diglossia implies a polarity of status between a low register dialect (L), with low s values, and a high register literary or official variety (H), with higher s values. Initially, diglossia was applied to dialect continua such as Demotics (L) versus Katharevussa (H) for Greek; the Marrocan or Tunisian dialect versus Classical Arabic in the Maghreb, or the s.c. basilect versus acrolect in Haitian Creole. Consequently, the term refers to a clear-cut dichotomy of social and symbolic functions according to formal (f) versus informal (i) settings: the L variety being used at home or with close friends, as a vernacular, whereas the H variety would be used at work and in public administration, so that in terms of status, H implies (f) and L implies (i) settings. Later on, Joshua Fishman and other sociolinguists working on migrant linguistic minorities in the USA extended the term beyond the restricted range of dialect continua, for example, between English (H) and Spanish (L) (Fishman and Cooper 1971). The former type of diglossia (structural continuum diglossia) is now labeled Fergusonian diglossia, whereas the latter (structurally discontinuous diglossia) is called Fishmanian diglossia. Indeed, it would be more appropriate to speak of distributed bilingualism in the case of Fishmanian diglossia, in order to restrain the use of the term diglossia to situations of pragmatic variation in the use of registers within the same language or dialect network. This would be useful, for instance, especially in situations of emerging additive bilingualism, to distinguish between a Fishmanian diglossia that is evolving toward balanced bilingualism, while within the formerly L status language, a new polarity is emerging out of the process of standardization. This is the case for Basque in Spain, where Basque as a co-official language in the Euskadi Community (or Basque Autonomous Community, i.e., CAB) refers to Euskara batua or unified Basque (the new standard written variety of Basque artificially created in the 1960s on the basis of several dialects from both the French and the Spanish side), while the local oral varieties, called 'Euskalkiak', now compete with this newcomer according to Fergusonian patterns of diglossia (i.e., sociolinguistic setting evolves toward a situation with X as the standard Basque, X' as Euskalkia or Basque local or regional dialect, Y as Spanish, and Z incorporating an intricate repertoire with embedded Basque inner diglossia (i.e., Fergusonian diglossia), though close to dilalia (see definition explained later).

Diglossia with Bilingualism. Diglossia with bilingualism can be represented symbolically as $\delta \cap Z$ or $\delta \cup Z$, where δ stands for diglossia and Z for bilingualism: these are further subdimensions that can be explored, $\delta \cap Z$ being a situation where diglossia is embedded in bilingualism, while $\delta \cup Z$ expands diglossia into bilingualism, unifying these two terms. In general, diglossia with bilingualism refers to in-group communication.

Diglossia without Bilingualism. Diglossia without bilingualism can be represented as $\delta - Z$ and is a situation typical of a highly segregated background or history, as in Mexico and Guatemala, where native communities could have no access to the official language (Spanish). In the late 1960s, most of the indigenous communities described in the four volumes of *Los Indios de México*, by the writer and journalist Fernando Benitez (1967–1972), were in this situation. Even now, in Chiapas and Oaxaca or Guerrero, sectors of indigenous communities in rural villages follow this pattern (women and young children before scolarization). This situation resulted in the emergence of bilingual elites, mostly bilingual school masters, or strong political movements claiming cultural autonomy and agrarian reform, since the 1970s—Zapatismo, in 1994, is typically an outburst of an initially segregated situation of this kind, and the result of a sociocultural and political uprising accompanied by a switch to a $\delta \cap Z$ situation.

Bilingualism without Diglossia. This situation, that can be represented as $Z - \delta$, has an outstanding example in the autonomous province of Voïvodina, in Serbia Djordjević (2004): Hungarian, Slovak, Romanian, Ruthenian, and Croatian are taught in primary and secondary schools, providing conditions for balanced bilingualism. Although the use of these minority languages varies according to urban or rural settings and the size and availability of the sociolinguistic network of speakers—Hungarian enjoying a wider social scope than Ruthenian or Romanian, for instance—alloglottic communities do not face any segregation. Moreover, except Ruthenian, most of these languages happen to be national languages of neighboring nations (Hungary, Romania, Slovakia, Croatia). If properly administered with a substantial degree of cultural and political autonomy, such a multilingual situation in a cross-border area may give outstanding results in terms of interethnic and international integration.

Neither Bilingualism nor Diglossia. A situation represented as $\delta - Z$ of strong diversity correlated to strong compartmentation of microcommunities, as in Papua New Guinea and in some regions of Equatorial Africa, is theoretically possible, although some amount of diglossia has to be taken into account, in relation with national or vehicular languages.

Dilalia. The term 'dilalia', formed with the Greek prefix and root *di-* ('two') + *lalêin* ('chat', 'talk'), has emerged recently from intricate situations of bilingualism associated or not with bidialectalism, or other patterns of multilingual use (Berruto, 1989), such as, for example, the coexistence of Italian and German in Alto Adige/Südtirol in Bozen (Bolzano), or of Spanish and Gascon in Val d'Aran in Spain, or Spanish, Romansh, and German or Italian in the Grisons (Switzerland), Catalan and the Balear dialect in the Balear archipelago, etc. In this case, models should account not only for the existence of X, Y, and Z but also of W as a coexisting official or vehicular language of status equivalent to or close to Y and of intricacy of the repertoire: X' and Y' or W'. In Grisons, Switzerland, for instance, one can encounter a speaker handling a language X such as Romansh, Y as German, as well as W as Italian, while Z would be a complex state of multilingualism with intricate patterns of dilalia: X' as local Romansh dialects Surselvan, Sutselvan, Surmeiran, Putèr and Vallader, Y' for Switzerland German, and even W' as the Ticinese Lombard dialects for those who had an opportunity to learn it moving to the Italian-speaking area.

As language proficiency is a cognitive asset, the complexity of repertoires may often be striking; the Vaupé situation in Brazilian Amazonia being one of the most famous cases of this potential intricacy (see Sørensen [1967]; Stenzel [2005]). Berruto (2004) suggests the matrix in Table 5.1 opposing the three main concepts at stake here, diglossia, dilalia, and bidialectalism.

Table 5.1 Caetano Berruto's typology of linguistic repertoires (Berruto, 2004).

		Diglossia	Dilalia	Bidialectalism
α	Sensitive to structural discrepancy between systems H and L	yes	yes	no
β	Use of both languages or varieties, X and Y, in casual talk and colloquial situations	no	yes	no
γ	X as the language of primary education and home	yes	no	no
δ	Clear-cut functional differentiation between X and Y	yes	yes	no (?)

Repertoires. The concept of 'repertoire' describes more complex, yet realistic, common situations, in which an individual uses more languages. In fact, a repertoire, in terms of language, is a complex cognitive system that can be defined as an adaptive multifold learning system (AMLS). In other words, an individual can more or less easily learn up to six or even ten languages within a lifespan, even though his/her command will be more balanced in two or three languages only. The languages or varieties of these languages learned by a speaker make up a repertoire—an array of languages or dialects one may rely on for *active* or *passive* communication. Although such multilingualism embedded in repertoires as a diversity of registers may sound queer from a European or a North American point of view, situations of intricate and multifold repertoires are usual in many African countries, or in some areas of South America. Under such circumstances, everyone is expected to speak a handful of languages, and often near to perfection, for specific social, economic, or symbolic purposes. If the fine-grained continuum between modalities of proficiencies (speaking and writing, resorting to *active skill* versus understanding and reading, amounting to *passive skill*) is taken into account, this phenomenon seems more familiar, or at least less dazzling.

Reactivated Proficiency. The phenomenon of 'reactivated proficiency' has to be taken into account especially in the case of endangered language preservation and revitalization; it mitigates, to a certain extent, 'language death': for example, a potential speaker, who has heard the language being spoken in his/her youth, even if not addressed directly in the lower status X language, may reactivate his passive proficiency and from this basis, develop autonomous active proficiency. Nowadays, this can be observed within the scope of Vepsian revitalization in the Republic of Carelia, in the Russian Federation, or with the endangered Huave varieties such as San Dionisio del Mar, in South-Western Mexico.[1] Indeed, reactivated proficiency is a main asset in all revitalization processes of endangered languages, as it allows one to take over the endangered language X after a consistent X → {Z + Y} process has taken place, in which the bilingual Z state melts into the final Y monolingual situation, still embedding Z properties awaiting conditions for X retrieval.

Revitalization. An endangered language can be prevented from undergoing language death or complete attrition through a social and cultural revival, eventually becoming resilient and starting a new social life. The process is quite intricate and relies highly on strong local will and empowerment, especially in education and everyday life. For this reason, although biological metaphors such as those used earlier are useful to describe these processes, one should possibly avoid such concepts as 'language death' or 'revival'. In fact, the main issue at stake here is to (re-)activate a {Z + Y} → Z process, which is socially and, to some extent, economically very demanding, especially at a broad scale, as it happens in some areas of the Basque country (the Navarre autonomous region in Spain, especially in the northern area) (Fishman, 1991, 2001).

Language shift. Basically, a 'language shift' is the disappearance of X monolinguals due to the presence of another language Y, following a competition process X → Z → Y which may be still active, eventually leading to a society composed of only Y monolinguals. Fishman (1991) proposed a standard of scale, spanning 8 stages, to be used for reversing language shift, amounting to a model of sociolinguistic resilience and revival.

- Stage 8: X is fading away and is only spoken by a few elders as an obsolete vernacular
- Stage 7: A few middle-aged speakers still have a more or less good command of X.
- Stage 6: X is still taught or spoken to children at home, or within the neighborhood.
- Stage 5: X is being taken into account in schools and education; alphabetization programs for children and adults in this language are implemented, either as private or public initiatives.
- Stage 4: Oral proficiency and alphabetization becomes widely spread through all generations and social groups.
- Stage 3: X expands in the work sphere; it is more and more widely accepted (i.e., tolerated), understood and used, and takes stronghold in official settings and in the media.
- Stage 2: XZ (the whole group of X speakers) is strengthening in every respect and the use of X has been generalized even in upper education and all areas of public services.
- Stage 1: X reaches equality of status with Y, and Z has generalized, or has gained popularity, which fosters X acquisition further.

Although this scale might seem idealistic, this is the aim of the linguistic policy initiated in Spain for languages such as Basque and Catalan since the 1980s.

The following is another scale for sociolinguistic revitalization or normalization converted to values of the status index $s \in (0, 1)$ (which is accompanied by an etiquette providing the

---

[1] http://axe7.labex-efl.org/node/329.

corresponding sociolinguistic correlates), based on situations observed in Mexico and in the Americas.

- $s < 0.10$: **Extinction**
- $s = 0.10$: **Obsolescence**. Vernacularity and obsolescence, very low status, no literacy (only some scarce and incomplete studies in the outer fringes of the local society). A high degree of acculturation in many domains of the sociocultural sphere;
- $s = 0.20$: **Residual Colloquiality**. Vernacularity and colloquiality among mid and elder generations, no transmission to youngsters, resilience with some kind of covert prestige fostering linguistic loyalty.
- $s = 0.30$: **Local Endemicity**. Vernacularity, with some active and endemic transmission to children at home (according to local and regional patterns of endemicity of sociolinguistic resilience and local prestige). Resilient sociocultural patterns in feasts and patterns of socioeconomic interactions within the region.
- $s = 0.40$: **Local Resilience**. Vernacularity, wide intergenerational transmission, some attempts at elaborating literacy in the language, a somewhat overt pride at speaking the language. Sociocultural cohesion, especially overt during intercommunal festivities.
- $s = 0.50$: **Incipient Mobilization**. Valorization of the language out of vernacularity, incipient literacy in some specialized sectors, especially in primary and secondary education, overt pride and linguistic and cultural loyalty. Sociopolitical mobilization to some extent, and active elaboration of local solutions to codification and standardization.
- $s = 0.60$: **Incipient Activism**. A rather wide use of the language, and overt local and regional identity, expanding use and prestige of the language from vernacular settings to intercommunitary agoras and to educative settings (primary and secondary schools); sociocultural and sociopolitical activism.
- $s = 0.70$: **Incipient normalization**. Intense revalorization and cultivation of the language in all spheres of local and regional life; strengthening teaching at school, not only as a specialized segment of the curriculum, but as a teaching language as well. Stronghold in medias, and substantial steps toward stable codification and standardization. Incipient YZ bilingualism.
- $s = 0.80$: **Expanding Normalization**. Expanding to many areas of the work sphere and in public institutions, expanding YZ bilingualism.
- $s = 0.90$: **Emerging Bilingualism**. Generalizing XZ, strengthening ZY.
- $s = 1.00$: **Additive Bilingualism**. Balanced bilingualism in which both languages X and Y are considered on the same footing by the Z bilingual community.

Indeed, one should bear in mind that these processes represent abstractions and that in real sociolinguistic frameworks, a linguistic community does not always evolve according to the same paradigm: these processes are prone to fluctuate or mingle into holistic states, possibly with some segments of the social continuum experiencing changes from one degree to the next, whereas others may indulge in recessive processes. The same could be said of many processes in the sociocultural or in the political sphere, as can be seen in stabilization or peace enforcing processes after a conflict; a situation that could be observed in Guatemala in 2000–2010, as far as democratization (or sociolinguistic revitalization) was concerned after a 35-year process of inner conflict. For this reason, we take over the notion of normalization by Aracil, Vallerdú, and Ninyoles that any situation of unequal bilingualism (or subtractive bilingualism) can be viewed as a conflict between competing sociocultural models through language. Any solution to conflicts of powers can be achieved through a normalization process. Of course, we are quite aware that the notion of 'languages in conflict' entails some reductionism, and we do agree with George Simmel that conflicts do not only sum up into a struggle and a win or lose competition, but also encapsulates intricate patterns of negotiation, compromise, cooperation, and mutual collusion—in short, conflicts make society possible. Moreover, normalization may take many different shapes, for example, expanding literacy is a kind of normalization, though it does not guarantee that the status of X is really definitely raised, or that further vitality of the oral variety or of the X' dialectal complex is safe. A highly endangered language can rise itself to the level of stage 0.90 and still remain more virtual than actual; the society may even turn out to be monolingual, as it is the case to some extent in Ireland. However, we will rather concentrate on less exceptional situations in this essay. For instance, the s scale discussed earlier in the range $(1, 10)$ here fits the situation of the majority of languages spoken today in Mexico and Middle America.

Vernacular. In Latin, this term initially referred to the 'domain of slaves'; it is used for low functions of a language, within a diglossic framework, when X is spoken only at home and hardly elsewhere, except the nearest neighborhood, especially in rural and socioeconomically depressed settings (in the case of indigenous languages of Middle America, for instance). In European settings, vernacular is more neutrally used for any L variety of X or Y, mostly spoken in the household, with informal status.

Vehicular. In the sociology of language, this term may refer to the wide use of a language, out of home and informal settings, as opposed to vernacular. Moreover, more widely in macro-sociolinguistics, it may refer to a lingua franca, or any language used for a broader scope of communication, as Swahili in Eastern Africa, or English, Spanish, or French in large areas of Africa and America.

Euskalkia. A name to indicate a generic Basque dialect, such as Navarrese, Gipuzkoan, Bizkaian, etc.

Erdara. In Basque sociolinguistic settings, this term, formed with the root *erdi-* ('middle') and suffix *-ara* for 'language' (e.g., *eusk-ara* 'Basque') refers to the main official language, as a vehicular language, on both sides of the Pyrenean, in Spain and France, that is, Spanish and French.

Chapter 6
Language Evolution Models

All language dynamics models try to describe language change in one way or another. There is no model that provides an exhaustive description of the dynamics of language, but each model tries to capture some relevant aspects. Any particular partitioning of language dynamics models into various categories is mainly a matter of convenience, as will become clear in the following sections. Language change can concern very different features, from morphological and phonological to semiotic and cognitive ones; or it can be observed as a language shift, a drastic form of language change in which a speaker adopts a new language in place of the former language; the two languages may not change appreciably during the shift.

6.1 Semiotic Dynamics models

Semiotic dynamics models (Castellano et al., 2009) owe their name to the fact that these models study the change of (the semiotic side of) a language, which is modeled as a set of couplings between words and objects (Chandler, 1994, 2007). During the interactions among agents, some word–object couplings can be turned on or removed, leading to the appearance of new languages. Semiotic dynamics models study how a unique language can emerge within a group of interacting individuals initially speaking different languages (Hurford, 1989; Oliphant and Batali, 1996; Castellano et al., 2009); the focus is on the mechanisms underlying the appearance of a consensus (Baronchelli, 2018). These models study neither the morphological changes nor the general cognitive changes, both names and objects being considered as fixed entities, but how the links between them can change in time.

6.1.1 The Nowak model

As the first example of a semiotic model, we consider the basic version of the Nowak model, which is an individual-based evolutionary language game similar to Hurford's model (not described here, see see Hurford [1989]), which has been used to tackle different questions. The mechanism leading to consensus about a certain language used by a whole group of speakers is based on the reproductive success assigned to an individual who has used a more successful communication strategy; eventually, only good communicators remain in the population; in the long time limit, the best communication strategy in the system is selected. The dynamical processes defining the model are (a) the interactions (communications) among individuals across the population, (b) the learning process from parent to offspring propagating a strategy from older to newer generations, and (c) population dynamics with the mentioned reproductive process. The main goal of the model is to find the best coupling strategy emerging from the interactions between agents.

In the basic scheme of the model (Nowak et al., 1999b; Nowak, 2000; Trapa and Nowak, 2000), there is a population of N agents, each one characterized by a 'language' defined by the probabilities that N_P known words are linked to N_Q observable objects: the generic nth individual is characterized by a matrix $P^{(n)} = \{p_{ij}^{(n)}\}$, whose elements represent the probabilities $p_{ij}^{(n)}$ that the nth agent uses the jth word ($j = 1, \ldots, N_Q$) to indicate the ith object ($i = 1, \ldots, N_P$), when in the role of the speaker; and by an analogous matrix $Q^{(n)} = \{q_{ij}^{(n)}\}$, whose elements represent the probabilities $q_{ij}^{(n)}$ that the nth agent interprets the uttered jth word as indicating the ith object, when in the role of the hearer.

The time evolution of the system proceeds through the following steps:

- Two agents n_1 and n_2 are picked randomly in the population to communicate with each other (in the original work, a swap over all units was used in place of random extractions), agent n_1 in the role of speaker and agent n_2 in the role of hearer.
- Agent n_1 randomly selects an object i and utters the jth word with probability $p_{ij}^{(n_1)}$. The probability that agent n_2 correctly interprets the word is the product of probabilities $p_{ij}^{(n_1)} q_{ij}^{(n_2)}$.
- After many communications between agents n_1 and n_2, the total probability of a successful communication is the average over all the words and all the objects available, $F(n_1, n_2) = \sum_{i=1}^{N_P} \sum_{j=1}^{N_Q} p_{ij}^{(n_1)} q_{ij}^{(n_2)}$. Averaging on the roles of speaker and hearer of the two agents gives the payoff function

$$F(n_1, n_2) = \frac{1}{2} \sum_{i=1}^{N_P} \sum_{j=1}^{N_Q} \left[p_{ij}^{(n_1)} q_{ij}^{(n_2)} + p_{ij}^{(n_2)} q_{ij}^{(n_1)} \right]. \tag{6.1}$$

This function, summed over all the other agents n_2 ($n_2 \neq n_1$), is assumed to provide the payoff function $\bar{F}(n_1) = \sum_{n_2} F(n_1, n_2)$ that determines the reproduction probability of individual n_1.

- In case of reproduction, the offspring adopts a language equal or similar to that of the parent.

Numerical experiments, which can differ in the details from that sketched here, provide some noteworthy results. *A priori* there is no reason why one should not expect that a word is coupled to many meanings or a meaning to many words since all the possible couplings can be realized by setting the matrix elements of a language different from zero. Instead, for example, in the case of square matrices (N' words coupled to N' concepts), the optimal communication strategy is realized when *each object is associated to only one word and vice versa*, corresponding to the matrices P and Q having only a value equal to one on each row and each column, while all the other elements are equal to zero, and to P being the matrix obtained by transposing Q.

It should be noted that this model focuses on the strategies of individuals, which can vary with time by switching on and off the couplings between words and objects depending on the outcome of the interactions. On the other hand, words, as well as the corresponding concepts, are given and are assumed to be fixed entities. In this sense, these models resemble mathematical models of competing quasi-species or chemical kinetics models. In fact, they can be formulated as competition models that describe the system in terms of the sizes of the populations of individuals with a given language (Komarova and Nowak, 2001a; Castellano et al., 2009).

There are various additional studies, generalizations, and applications of this model; many are not considered here. Some variants include noise in communications (Nowak and Krakauer, 1999; Nowak et al., 1999a) and extend the application range to other problems such as language acquisition (Komarova and Nowak, 2001b) and syntax development (Plotkin and Nowak, 2001).

A phase transition is observed, in which the dominance of one learned language is replaced by the coexistence of many languages, as the parameters are varied (Nowak et al. [2001, 2002]). This result is relevant for the study of language size distribution (Section 6.2.2).

A spatial variant of the Nowak model was introduced and analyzed by Di Chio and Di Chio (2006, 2007, 2009), who studied the influence of the topology of an underlying spatial environment on the model dynamics, when the fitness function and the number of offspring explicitly depend on the distances between individuals; these effects make the system self-organize into a wide spectrum of different equilibrium configurations, spontaneously emerging from random initial conditions, possibly containing mono- or multilingual clusters—for further details, see Di Chio and Di Chio (2009).

6.1.2 Naming game models

From a semiotic dynamics viewpoint, the basic version of the naming game (NG) model (Baronchelli et al., 2006b; Baronchelli, 2016) resembles a single-object version of the Nowak model, in that several individuals are competing with each other in order to assign different names to a single object. However, the dynamics of the NG model is different. In its basic version, there is neither population dynamics nor reproductive advantage: the number of agents is conserved and consensus can be reached through a negotiation process, as described below.

In this section, we will discuss the main features of the basic model, in which the system dynamics for an N-agent system are defined by the following update rules (Baronchelli, 2016).

(i) Each agent is equipped with an inventory, where the words learned by the agent are recorded. Initially, the inventories of all agents are empty.

(ii) At a new time step, two agents are randomly selected, one of them acting as speaker, the other one as hearer.

(iii) The speaker randomly extracts a word from the inventory (or invents a new one if the inventory is empty) and conveys the word to the hearer; this process can be understood as the speaker utters the name while pointing at the corresponding external object, so that

the hearer records not just the name but also its association to the object. At this point, two different situations can take place.
   (a) Communication success: The conveyed word is present also in the hearer's inventory; the two agents erase the other words in their inventories, leaving only the conveyed word.
   (b) Communication failure: The word conveyed is *not* present in the hearer's inventory; the hearer records the word, adding it to the inventory.

Examples of an unsuccessful and a successful communication are schematized in Figure 6.1. Despite its simple structure, the NG model has the remarkable property of reproducing the spontaneous emergence of consensus (about which word is used). The model follows a set of simple interaction rules that are symmetrical in the words used; it does not follow any central control (Baronchelli et al., 2007).

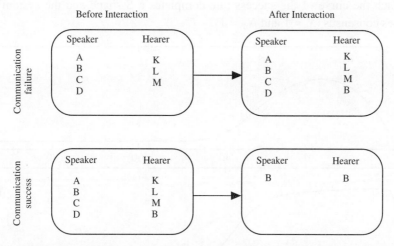

Figure 6.1 Two examples of pair-wise interactions in the basic naming game model, in which the speaker conveys word B to the hearer. Left (right) panels represent the initial (final) states of the agents, before (after) the interaction. Top panels: Communication failure, the hearer does not have the conveyed word B and adds it to the inventory. Bottom panels: Communication success, the hearer already has the conveyed word B and the two agents reset the inventories by keeping only B and deleting the other words.

The time evolution of the system can be effectively visualized through some key quantities such as the total number of words N_w (which represents the total memory of the system), the number of *different* words N_d (representing the diversity in the system), and the rate S of successful interactions between agents. These three quantities, averaged over 1200 simulations, are plotted in panels (a), (b), and (c), respectively, in Figure 6.2 in a time interval of 10^5 time steps, for a basic NG model with $N = 1000$ agents (Marchetti et al., 2020). In particular, panel (b) shows the system spontaneously undergoing a (disorder/order) transition toward an asymptotic absorbing state, where a word takes over all the others. It can be shown that this transition always takes place in the minimal NG (Baronchelli et al., 2006a).

Figure 6.2 suggests that the non-equilibrium dynamics presents three different characteristic regimes.
(i) In the first phase, most of interactions fail; agents invent new words and store them in their own inventories. As a result of the uncorrelated games, N_w and N_d increase linearly with time ($N_w \simeq 2t$, $N_d \simeq t$) and the success probability is still very small, $S(t) \approx 3t/N^2$. The number of different words created by the agents during this phase reaches a value of the order of $N_d \simeq N/2$.
(ii) Then the system reaches the second phase characterized by the plateau of $N_d(t)$, visible in the curve depicted in panel (b) of Figure 6.2. The $N_d \simeq N/2$ created words spread through the whole system and the first relevant correlations between words and agents appear due to the overlap between the agents' inventories.
(iii) At this point, the number of different words begins to decrease and the third phase begins, in which the curve of the success rate completes a S-shape and the system eventually reaches consensus ($S = 1$ and $N_d = 1$).

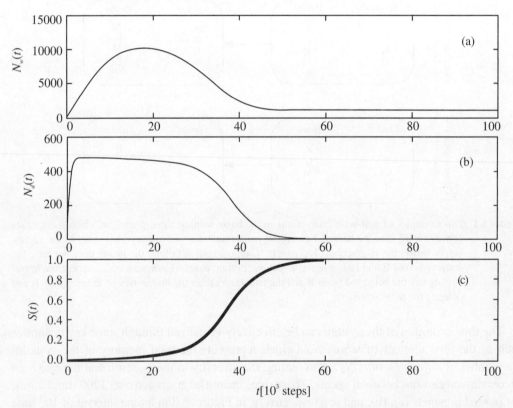

Figure 6.2 Average values of N_w, N_d, and S over 1200 simulations of 10^5 time steps for a population of $N = 1000$ agents in a mean field approach. From Marchetti et al. (2020).

The convergence to a consensus state is sudden, a feature that becomes evident by studying how the shape of the success rate S versus time scales with the population size N (Baronchelli et al., 2006a), as shown in Figure 6.3. If t_{max} represents the time at which N_w reaches its maximum value $N_w^{\max}$ and t_{conv} is the convergence time, at which consensus is reached ($N_d = 1$), then one finds that

$$t_{\text{conv}} \sim t_{\max} \sim N^\alpha, \qquad (6.2)$$

with $\alpha \simeq 1.5$, a value that can be derived either on an analytical basis or from simulations (Baronchelli et al., 2006a, 2008). The maximum amount of memory required (the cognitive effort) by agents grows with N with the same power law, $N_w^{\max} \sim N^{1.5}$.

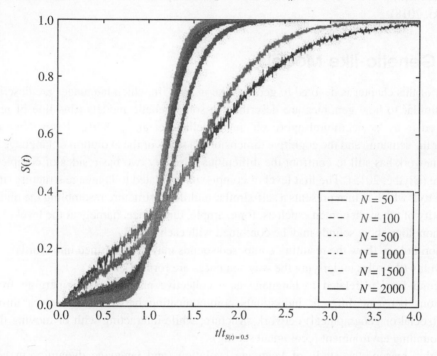

Figure 6.3 Average values of success rate S versus the rescaled time $t/t_{S(t)=0.5}$ for system sizes $N = 50, 100, 500, 1000, 1500, 2000$ obtained from 1200 runs. Increasing the size, the curves become steeper, implying a faster convergence to consensus. From Marchetti et al. (2020).

Besides the basic version, various extended versions of the NG model have been introduced in order to study alternative, more realistic scenarios, for example, those including noise in communication or involving more than two agents at time; or embedding the inter-agent dynamics on a complex network topology; all these extended versions maintain some form of the negotiation process (Chen and Lou, 2010). While in the basic version, inventories can be arbitrarily large, the Q-conventions NG models have a vocabulary size limited to a finite number Q of different conventions. The 2-conventions NG model will be considered in detail in Chapter 7 as a competition model. For a review of various versions, the reader is referred to Loreto et al. (2011); Chen and Lou (2010).

Overall, the NG has turned out to be a fruitful and general framework for the simulation of communications between individuals. The NG can also be promoted to an actual evolutionary model with agents characterized by an evolutionary selected learning ability (see Lipowska [2011]); it can be used as a model of interaction between language and biological evolution (Baldwin effect, see Lipowski and Lipowska [2008]), revealing phase transition when the communication skills reach a critical threshold. The NG can also be generalized to deal with P objects rather than with only one, thus retaking the complexity of the original Hurford model and the Nowak model, which allows the study of the evolution and interactions within a language of synonymy and homonimy, and/or promoting the basic negotiation dynamics to a reinforcement process (Lipowski and Lipowska, 2009; Lenaerts et al., 2005; Lipowska and Lipowski, 2018).

6.2 Genetic-like Models

The rest of this chapter is devoted to genetic-like models, in which languages are described in a way similar to how genomes are described in some genetic models (this line of research is referred to as 'sociocultural approach' in Castellano et al. [2009]). Even if we do not consider the semantic and the cognitive dimensions, a study of the evolution of language within this framework has still to confront the difficulties raised by two basic sides of complexity in language (Blythe, 2015). The first level of complexity is related to language structure (internal complexity), which in turn presents a self-similar multilevel structure resembling the multiscale complexity of other phenomena such as, for example, turbulence, namely at the level of:

- phonology (how sounds may be combined with each other);
- morphology (how the resulting composed sounds may be combined into words);
- syntax (the rules underlying the way sentences are formed).

The second level is related to language as a collective phenomenon, emerging from the interactions between different individuals communicating between themselves through a certain (social or geographical) network structure, while interacting with or moving through the surrounding environment (ecological setting).

For this reason, the study of language evolution (and language dynamics in general) is a particularly delicate subject since the various levels of complexity can interact with each other shaping the overall language evolution processes. It is clear that a satisfactory understanding and description of language dynamics has to go through an integration of all the different complexity levels, as emphasized, for example, by Shen (1997) in the case of sound change. The diffusion of a linguistic innovation has many analogies with other diffusion processes, in particular, the diffusion of social innovations (Rogers, 1983) or epidemic spread (Pastor-Satorras et al., 2015); however, it has its own specific complex features due to the fact that it takes place across the aforementioned dimensions which impose their specific constraints. Among the common features between linguistic innovation and other diffusion processes, we mention (a) the social structure which can usually be captured in terms of a complex network representing the connections between the units of the group and (b) an 'S-shaped' curve characterizing the dependence of the fraction of adopters of an innovation versus time.

6.2.1 Statistical properties of languages

Besides the huge amount of data about language structure, mutual intelligibility, etc., contained in linguistic databases, such as those considered in the case studies discussed in the previous chapters, there exists a set of empirical statistical properties of languages, when considering each language as a single unit of the system under study.

First, languages are characterized by a typical language size distribution (see e.g., Gomes et al. [1999]) that has an approximately log-normal shape, which can in turn be approximated by a power law in a wide range of the size (Wichmann, 2005).

Other properties of languages concern their geography and have analogs in the properties of biological species (see Villalobos and Rangel [2014]; Gomes et al. [1999]). They express mathematical relations connecting the size of languages (number of speakers), their diversity (number of different languages), and geographical properties, such as the area in which they are spoken (geographical extension) (Gomes et al., 1999; Sutherland, 2003; Solé et al., 2010). Detailed overviews of such properties can be found in Gomes et al. (1999); Capitán and Manrubia (2014).

A proper fitting of the language size distribution is one of the goals of some models of language dynamics, such as the model of Schulze and Stauffer that will be considered in Section 6.2.2. When considering geography-related properties, the model has to be embedded in a space with one or more space dimensions; some models consider space in the form of a lattice made up of discrete sites. For example, the diversity–area relation is considered in the Viviane De Oliveira model, discussed in Section 6.2.4; this model is embedded in a two-dimensional square lattice. A minimal type of geography, represented by a lattice, is also used in the bit-string evolution model presented by Kosmidis, Halley, and Argyrakis (see Section 6.2.3), which exhibits interesting phenomena of clustering.

Models of language dynamics (limited to a description of a competition process) containing geography as a basic element are discussed in detail in Chapter 7.

6.2.2 The Schultze and Stauffer Bit-String model

The bit-string model of Schulze and Stauffer (2005) is an individual-based model inspired by analogous models in biology. The model has been used to study various questions, from the fate of a language undergoing competition and evolutionary processes under the action of ecological constraints to the prediction of the observed shape of the language size distribution. The original goal of the model was to describe cultural spreading, including programming languages, sign languages, bird songs, alphabets, etc. The Schulze and Stauffer model can be generalized in many ways. It is instructive to see how these questions can be tackled through a simple version of the model, illustrated below.

In the first version of this model, there are N units (e.g., N speakers) and each unit n ($n = 1, \ldots, N$) has a 'language' represented by a string $b^{(n)}$ made up of Q bits, $b^{(n)} = \{b_q^{(n)}(t)\} = [b_1^{(n)}, b_2^{(n)}, \ldots, b_Q^{(n)}]$, that is, variables $b_q^{(n)}$ that can assume either the value 0 or 1 (originally, the model was studied for $Q = 8$ and $Q = 16$). The simplification of a language in terms of a string is not meant to realistically model the complexity of an actual language but to represent a linguistic trait through a minimal set of variables in order to focus on the mechanisms of the evolution process. The string can represent, for example, a morpheme (for instance, if the string has $Q = 10$ bits, the total number of possible combinations of 0 and 1 of the single bits are $M = 2^Q = 2^{10} = 1024$). The number $M = 2^Q$ represents the maximum number of different 'languages' possible in the system.

In the model, vertical transmission of a language from parent to offspring takes place. Furthermore, random mutations can randomly change the string of an offspring. The model also contains the Verhulst population dynamics with a reproduction (Malthus) rate r and carrying capacity K. The model evolution proceeds, using update rules, as follows.

Initial conditions. At $t = 0$, there is a single individual, $n = 0$, speaking the 'language 0', defined by the string $S = [0,0,0,0,0,0,0,0]$ for the case $N = 8$—this is just a conventional initial state.

Time evolution. At each time step, the following events take place:
  (i) Reproduction. Each individual produces (asexually) an offspring which inherits the same bit-string.
  (ii) Mutation. A change in a randomly chosen bit of the offspring's string can take place with probability p.
  (iii) Competition. There can be a language shift, a 'horizontal' change resulting from the direct competition between different languages. An individual n can adopt the *entire bit-string* of another agent n' with a probability proportional to $(2N/K)(1 - x_n^2)$, where x_n is the fraction of population speaking the same language of agent n. The factor $2N/K$ is introduced by the authors to make the probability of shift smaller at the beginning (when $N \ll K$) than at the end ($N \approx K$).
  (iv) Time is increased by one step and the process reiterates starting from point (i).

A main result is an order–disordered transition as parameters are varied, corresponding to two *qualitatively* different possible outcomes, observed in the numerical simulations (10^6 agents and 1000 time steps in the original work).

- For a small enough mutation rate p, the original language remains the most common one, spoken by nearly all individuals; the mutations represent minorities.
- For values of the mutation rate p above a threshold, many languages are generated and the original '0 language' is not the most common anymore—no language actually dominates. Instead, a roughly log-normal size distribution emerges similar to the actual size distribution of human languages (Sutherland, 2003; Wichmann, 2005), see Schulze and Stauffer (2005) and arxiv.org/abs/cond-mat/0411162.

An interesting point to note while approaching this problem through a mathematical model is that these two qualitatively different scenarios correspond to the very same model with different values of the mutation rate. In other words, the values of the parameters determine whether there will be a basically unique variety of languages or not.

A similar order–disorder phase transition is observed in the model presented by Nowak et al. (2001); this model shows a change from the dominance of one learned language to the coexistence of many different languages.

Schulze and Stauffer studied several variants of this model (see Stauffer and Schulze [2005]; Stauffer et al. [2006]; Schulze et al. [2008] for some generalizations, applications, and reviews). In particular, the bit-string model was generalized into a model in which the bit-strings $b^{(n)}$ are replaced by strings of variables, $s^{(n)} = \{s_q^{(n)}(t)\} = [s_1^{(n)}, s_2^{(n)}, \ldots, s_Q^{(n)}]$, in which the Q variables $s_q^{(n)}$, here representing linguistic traits, can assume F different values in the interval $s \in (1, F)$ (Schulze and Stauffer, 2006). Since each string models a different language, there can be Q^F different possible languages in the system.

A spatial version of the Schultze model was applied to the study of language evolution in a model system in which a barrier divides the area into two regions (Schulze and Stauffer, 2007). The dynamics is similar to that of the original model, with the difference that (1) the model is embedded in a lattice where the barrier limits the interactions between agents and each agent can only influence the neighbors; and (2) mutations change the value of a variable in the wider interval $(1, F)$, with $F > 1$. The dynamics depends on the parameters p and q, representing mutation rate and linguistic diffusion rate, that determine the final scenario. While in the absence of a barrier, the system eventually fragments into a large number of small groups speaking different languages, the presence of a barrier allows the emergence of an ordered state, in which two different languages dominate the two opposite sides.

Another spatial variant on the square lattice was studied by Teşileanu and Meyer-Ortmanns (2006), with the relevant new feature that the (dis)similarity between languages is taken into account: the Hamming distance between the languages of two agents (which coincides with the Levenshtein distance between two fixed-length strings) influences the rate at which the language of one agent can be adopted by the other one.

6.2.3 The Kosmidis, Halley, and Argyrakis Bit-String model

The model of Kosmidis et al. (2005) provides another nice example of how different processes such as competition and evolution, population dynamics, and diffusion, can be merged into a single multi-process model. In this model, there are N individuals interacting with each other. Each individual n ($n = 1, \ldots, N$) can have a genetic label $\mathcal{A}$ or $\mathcal{B}$ that is fixed in time, and an associated language defined by a string $s^{(n)}$ of $2Q$ bits,

$$s^{(n)} = \{s_q^{(n)}\} = (s_1^{(n)}, s_2^{(n)}, \ldots, s_Q^{(n)} | s_{Q+1}^{(n)}, \ldots, s_{2Q}^{(n)}), \qquad (6.3)$$

which can change in time as individuals interact with each other, similar to the bit-string model of Schulze and Stauffer (Section 6.2.2). Each bit represents a linguistic trait known to individual n, if $s_q^{(n)} = 1$, or unknown if $s_q^{(n)} = 0$. There are initially only two languages in the system: language A defined by the first Q bits being all $s_q^{(n)} = 1$ (and the rest of the bits equal to zero) and language B defined by the second half of bits $s_q^{(n)} = 1$ (and the first half equal to zero). For example, for $Q = 4$, language A is $s_A = (1, 1, 1, 1|0, 0, 0, 0)$ and language B is $s_B = (0, 0, 0, 0|1, 1, 1, 1)$. The string with $s_{AB} = (1, 1, 1, 1|1, 1, 1, 1)$ (for $Q = 4$) defines a bilingual with perfect knowledge of both language A and B. The model dynamics is defined by the following update rules.

Initial conditions.
- The initial $N(t = 0) = N_0$ individuals are partitioned into two groups of N_{A0} and N_{B0} individuals, who have genetic label $\mathcal{A}$ and $\mathcal{B}$, respectively.
- Individuals $\mathcal{A}$ and $\mathcal{B}$ are assigned language A and B, respectively.
- All individuals n ($n = 1, \ldots, N$) are assigned the same fitness value $f_n \equiv f_0$.
- Each agent is randomly located on the site of an $L \times L$ square lattice — not more than one individual per site (originally $L = 100$).

Time evolution. At each time step t.
  (i) An individual i is randomly selected.
  (ii) A lattice site is randomly selected among the first neighbors of agent i.
  Two possible processes can take place:
   - *Diffusion.* If the selected site is empty, individual i jumps to that site.
   - *Interaction.* If the selected site is occupied by another individual j, a language communication takes place between i and j, as follows:
     – *Fitness update.* The fitness of i and j is updated, $f_i \to f_i + \Delta f$, $f_j \to f_j + \Delta f$, with Δf defined as their number of common bits $\Delta f = \sum_{q=1}^{Q} s_q^{(i)} \times s_q^{(j)}$.
     – *Learning process.* With probability p_l, agent i learns a randomly chosen feature known by j and unknown to i (the corresponding bit is set from 0 to 1); the same takes place for the jth individual with a randomly chosen feature known by i and unknown to j.
     – *Forgetting process.* With probability p_f, individual i forgets a randomly chosen known feature (turning the bit from 1 to 0). The same takes place for the jth individual, with a feature known by j.
  (iii) *Reproduction.* A neighbor site of agent i is randomly selected and if that site is empty, individual i reproduces with probability p_r proportional to the fitness f_i and the offspring is located in that site, inheriting genetic label, language, and a reduced fraction of the fitness of the parent i.[1]
  (iv) *Death.* With probability p_d, individual i dies and is removed from the lattice.
  (v) Time is incremented by one time step and the loop is reiterated.

The authors carried out many analyses of the model, studying in particular the coexistence probability for the two populations $\mathcal{A}$ and $\mathcal{B}$. Among the various results, they found that the interplay of the various time scales associated with the various dynamical processes leads to a complex time evolution of the system, in which a stable state emerges, where all the individuals (both of the $\mathcal{A}$ and $\mathcal{B}$ population) are bilinguals—in other words, the system self-organizes toward a bilingual state. In the long time scale, one of the two populations (either $\mathcal{A}$ or $\mathcal{B}$) eventually disappears; this always happens, even though, initially both $\mathcal{A}$ and $\mathcal{B}$ are symmetrical in population size, string length, and fitness. From a historical point of view, the model describes how the language of a population that has disappeared can leave relevant traces in the language of the surviving population.

---
[1] There are two versions of the model, a global version, considering the fitness of each individual in the lattice, and a local version, in which only the neighbors of individual i are taken into account. See Kosmidis et al. (2005) for details.

The model exhibits interesting spatial structures due to a segregation process caused by the interplay between diffusion and language spreading, in which individuals of a certain genetic type are located more probably near individuals of the same type, in spite of the initial random conditions and the interactions.

6.2.4 The Viviane De Oliveira model

The model of De Oliveira et al. (2006b) provides a minimal description of a population reproducing and expanding across a territory, while its language changes and splits into different languages. Geographical expansion will be considered again in detail in the next chapter.

The De Oliveira model is an individual-based model, in which agents can be interpreted as speakers or groups of speakers, for instance, villages or farms.

'Languages' are defined in a minimal way by simply counting them (no structural property is described) through a counter n running from $n = 1$ (the language of the ancestral group) to the number of languages in the system at time t, $n = D(t)$, referred to as *language diversity*. In other words, languages can split, generating new languages, but are otherwise fixed entities between two splitting events.

There are two relevant parameters in the system dynamics. The first parameter is the local fitness: the territory is assumed to be inhomogeneous; it is represented as a square lattice made up of $A = L \times L$ sites, with a different local *carrying capacity* $C(\ell) \in (0, 1)$ for each site, representing the local population sustainability. Furthermore, a site ℓ, occupied by an agent i, has a *social linguistic fitness* $f(\ell)$ given by the sum of the carrying capacities $C(\ell')$ of all the other sites ℓ', in which the *same* language of site ℓ is spoken: $f(\ell) = \sum_{\ell'} C(\ell')$.

Despite its minimal framework, this model can predict some statistical properties of languages related to their spatial distribution. The details of the model are illustrated as follows.

Initial conditions.
- All lattice sites are empty, but one: In a randomly extracted site ℓ_0, the first individual of the system (the 'ancestral group') is placed.
- Each site ℓ ($\ell = 1, \ldots, A$) of the inhomogeneous territory is assigned a local *carrying capacity* $C(\ell)$ extracted randomly in the interval $C \in (0, 1)$.

Time evolution.
(i) With probability $p \propto C(\ell_1)$, the next site ℓ_1 that will be colonized is randomly extracted among the neighboring unoccupied sites.

(ii) Agent i that will colonize site ℓ_1 is randomly extracted among the agents living in the neighboring sites of ℓ_1.

(iii) The selected agent i reproduces and the offspring j is located at site ℓ_1, adopting the same language of the parent i.

(iv) With probability $p_m = \alpha/f(\ell_1)$, where α is a constant and $f(\ell_1)$ the social fitness of the colonized site ℓ_1:

- the language of the offspring j undergoes a mutation and becomes a new language that is assigned the new counter value $n = D + 1$.
- diversity is updated, $D \to D + 1$.

(v) Time is updated and the loop restarts from point (i)—until all the cells of the lattice have been colonized and then the simulation stops.

A typical geographical 'star-shaped' configuration obtained in this way is shown in Figure 6.4. In fact, such star-shaped configurations are observed in some linguistic landscapes (see, for example, Section 4.3 on the Numic language).

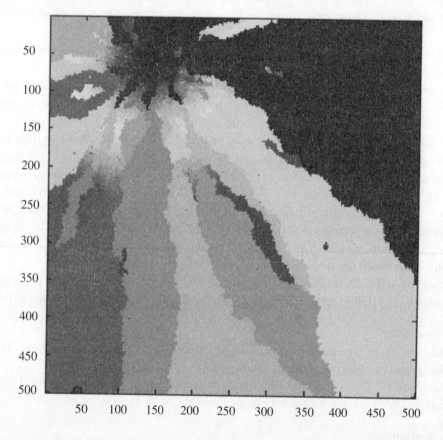

Figure 6.4 Star-shaped geographical distribution of languages. Reprinted from De Oliveira et al. (2006b) with permission from Elsevier.

What is even more interesting is the area–diversity relation (i.e., between the lattice area A and the final diversity D) shown in Figure 6.5: numerical results are consistent with the experimental law $D \propto A^z$ found in Gomes et al. (1999), with an exponent (of the first part of the curve) $z \approx 0.43 \pm 0.02$ for $\alpha = 0.73$ (upper curve) and $z \approx 0.88 \pm 0.01$ for $\alpha = 0.3$

(lower curve). Furthermore, if $A_i(t)$ is the area of the ith language at time t, the average area covered by languages at time t is

$$\bar{A}(t) = D^{-1} \sum_{i=1}^{D(t)} A_i(t).$$

The authors found that $\bar{A}(t) \propto t^\beta$, where $\beta = 1 - z < 1$. Comparison with standard Brownian diffusion of an initial spot, characterized by an area $A(t) \propto t$, shows that languages undergo a subdiffusion process (Ebeling and Sokolov, 2005). The same factor β appears in the relations of the average area of a language versus the side L of the lattice employed, which shows that the language area grows less than expected with side L (or with the simulation area), $\bar{A}(L) \propto L^{2\beta}$. Further details are given in De Oliveira et al. (2006b).

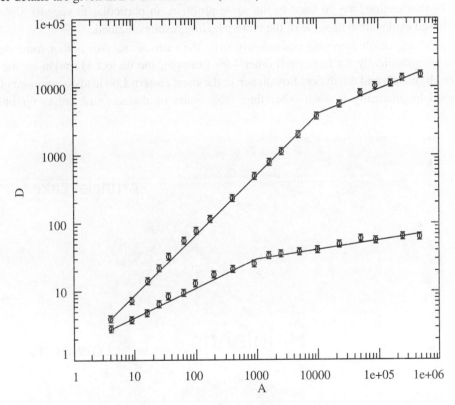

Figure 6.5 Area–diversity relation from the numerical experiments of De Oliveira et al. (2006b). Reprinted from De Oliveira et al. (2006b) with permission from Elsevier.

The authors also studied some generalizations of this model (De Oliveira et al., 2006a), in which, for example, language substitution (colonization of occupied sites) was possible and the fitness of a language saturates when reaching a randomly assigned threshold. In this case, agreement with known data (Gomes et al., 1999) was found concerning both a double power-law regime of the language diversity distribution and the log-normal shape of the language size distribution, though a strong sensitivity of results on the model assumptions was pointed out.

A noteworthy development of this model is its merging with the string structure of the model of Schulze and Stauffer discussed in Section 6.2.2 and its application to the study of language size distribution. Replacing the counter n with a string structure and the uniform local capacity distribution with a more realistic distribution provide a better and more robust language size distribution similar to the real one (De Oliveira et al., 2007; De Oliveira et al., 2008).

6.2.5 Modeling the Mazatec expansion

The Mazatec dialects were introduced and discussed in Section 3.5, where a study of the corresponding dialect network was carried out through various algorithmic and visualization tools. In this section, we go back to the same problem, in particular to suggest a possible model-based explanation of some of the more puzzling results obtained.

A surprising result emerging consistently from the various analyses is that some dialects that are geographically far from each other—for example, the dialect Mazatlán on the most western Highlands and the dialect Soyaltepec in the most eastern Lowlands—are more closely connected linguistically to each other than other pairs of dialects that are geographically

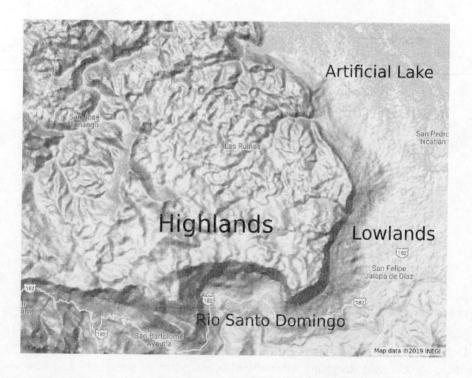

Figure 6.6 Enlarged portion of a terrain map of the Mazatec area showing the presence of small valleys connecting the artificial lake in the Lowlands to the Highlands. (Map data © 2019 Google, INEGI.)

close to each other, such as the dialects Jalapa and Ixcatlān in the southern Lowlands, see the geographically embedded Levenshtein distanceand mutual intelligibility dialect networks shown in Figures 3.8 and 3.9. In fact, both a large geographical distance or the presence of mountains alone would be sufficient reasons to expect a higher—rather than a lower—linguistic distance. These two different aspects of the same puzzle turn out to be related to each other and to have a simple explanation when a direct analysis of the spreading process is carried out through a numerical simulation.

Nevertheless, a careful inspection of the terrain map of the Mazatec area (Figure 6.6) reveals that besides the main valley of Rio Santo Domingo, a natural entrance to the Lowlands when coming from the West, there are smaller valleys, rising from East to West, that is, from the Lowlands to the Highlands, which can serve as entrances to the Highlands. In order to simulate the Mazatec spreading process, we set up a very simple numerical experiment based on the geographical toy model shown in Figure 6.7: a square lattice is used as a scheme of the Mazatec region. Crosses represent the corners of the lattice sites, while grey areas are inaccessible regions representing the higher mountains. The region labeled as 'valley 1' stands for the valley of Rio Santo Domingo. The corresponding entrance in the valley (and in the Mazatec region) is explicitly shown with a big arrow. This is where the Mazatec spreading process is thought to have began, when the people speaking the Mazatec proto-language entered the valley of the river Rio Santo Domingo along the West→East direction to diffuse in to the Lowlands and eventually across the Highlands. The area on the right-hand side models the wide region of the Lowlands; finally, the narrow channel on the top, labeled 'valley 2', represents one of the small valleys mentioned visible in the terrain map in Figure 6.6.

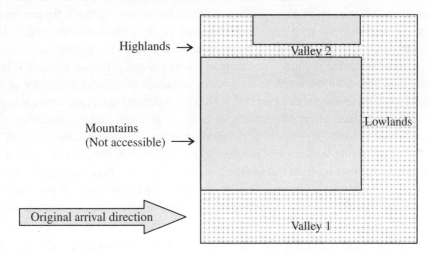

Figure 6.7 Two-dimensional lattice used in the simulation of the Mazatec expansion.

The model evolves with time following the dynamics of the DO model (see Section 6.2.4). In the initial conditions, the first 'ancestral group' entering the Mazatec region is located in the bottom-left site in Figure 6.7. The model used is actually a version simplified in various aspects, for example, the terrain is assumed to be homogeneous, apart from the mountain regions in gray that are not accessible, and the fact that death processes are neglected, that is, each time a site is colonized by an agent speaking a certain language, it will remain as such. At each time step, agents reproduce and spread across the empty lattice sites, while languages possibly undergo language innovations that create new languages, interpreted here as the appearance of a new Mazatec dialects.

This model is clearly an oversimplification of many relevant aspects of the complexity of language evolution that may be taken into account, but the point is to check whether the interplay of mutations and diffusion only can provide some explanations. There are two relevant time-parameters in the model, the growth rate r_g at which an agent reproduces and a new group is generated that occupies a neighbor (empty) site and the mutation rate r_m, at which a language splits into two different dialects—in fact, there is a single parameter since it is only the ratio of these two time scales that is relevant. In order to distinguish a new language from the older ones, we assign a new label value to it and, for the purpose of visualization, we plot the corresponding sites with a new color. The maps of an example of spreading process at different times, from time $t = 335$, when valley 1 has been partially colonized, to the final time $t = 1257$, when all the accessible sites of the simulation box have been colonized, are shown in Figure 6.8.

The snapshots in Figure 6.8 indicate that something peculiar happens during the population evolution. One can notice that about 450 time steps are needed between the entrance of the first agent into valley 1 (at the left-bottom site) and the arrival of the new generations to the lowlands, after having colonized valley 1. Then, it takes at least 600 time steps more for the agents to spread as far as the top of the simulation box, diffusing across the lowlands. However, it takes less than 200 time steps approximately to cross the narrow valley 2, for the simple fact that valley 2 is very narrow with respect to the other regions (lowlands and valley 1). This can be expressed by saying that valley 2 (at variance with valley 1) is a quasi-one-dimensional domain: its transverse size is much smaller than its length and diffusion through it proceeds much faster than across an open space, such as the lowlands. From the viewpoint of ecology and biogeography, this means that a corridor, which is by definition a place where it is possible to move more easily than along other paths, in order to go from a place to another place, is also a place where one moves *faster* if the corridor is narrow (quasi-one-dimensional), even if the type of the underlying landscape is similar. As a consequence, the Mazatec diffusion from the East to the West, that is, from the lowlands to the highlands, may have taken place much faster even than the diffusion across a smaller distance in the open area of the valley of Rio Santo Domingo or of the lowlands, implying in turn the appearance of a smaller number of innovations and a closer similarity with older dialects.[2]

The simulation shown in Figure 6.8 is obtained for an optimal ratio between the reproduction and the mutation rate, which best simulates the data about Mazatec, in particular the close similarity (here represented by the same color) between the western Highlands and the northern Lowlands dialects. Notice also that the final number of dialects is comparable with the actual one. Simulations carried out using higher mutations rates provide too many

---

[2] A video of the numerical simulation gives an idea of the different spreading speeds:
youtube.com/watch?v=s4_eKQPG-ag&list=PL3lmCtliK8NCu6dGLEv_url_9Y_91kFGg&index=2&t=0s.

different dialects; while smaller values of the ratio produce a final linguistic landscape that is too homogeneous with respect to the real one.

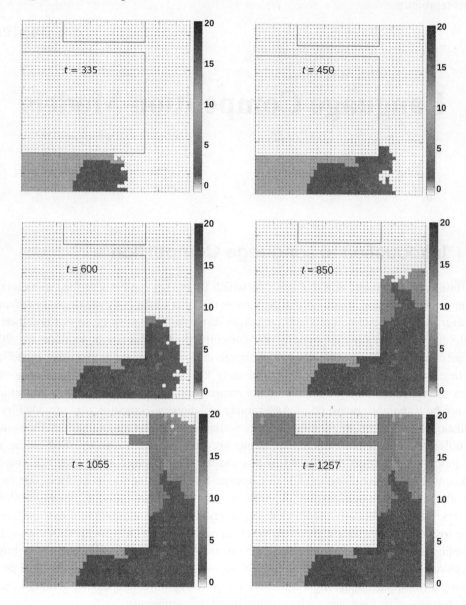

Figure 6.8 Time evolution of a simulation of the Mazatec expansion. Each different gray tonality corresponds to a different language.

Chapter 7

Language Competition Models

7.1 Introduction to Language Competition

The time scale of competition dynamics on which languages can be described as an analogy of competing biological species is usually shorter than that of the fully evolutionary dynamics. However, it is a time scale on which languages can either spread or disappear and therefore is relevant for the destiny of most of the currently existing languages (Solé et al., 2010). Competition models are technically more simple than evolutionary or cognitive models, but in a linguistic system, one first needs to identify the competitors and the main competition processes in order to describe them within a competition and natural selection paradigm; this may not be a straightforward task. An example of language competition is provided by two linguistic features competing with each other without undergoing major changes, for example, two different ways of pronouncing the same word or two synonyms referring to the same object/idea. The models considered in this chapter can be categorized either as two-state models, when there are monolinguals of language X and language Y, or as three-state models with bilinguals, where in addition to X and Y monolinguals, there is also a Z group of bilingual speakers. These models provide a simplified description of the adoption of a new language or of the loss of a known language as abrupt changes of the state of a speaker: X→Z or Y→Z (monolingual→bilingual) transitions and the inverse Z→X or Z→Y (bilingual→monolingual) transitions. The overall process X→Z→Y (Y→Z→X) represents a language shift, the process whereby a community speaking a certain language X (Y) shifts to speaking another language Y (X) because of its interaction with another linguistic community.

The paradigm of language shift is extensively studied and still represents a huge challenge for mathematical modelers. Usually, languages that are considered to be more prestigious expand at the expense of other languages. However, there are many possible causes behind a language shift. From a historical perspective, some questions have remained unanswered, puzzling linguists. These questions can receive at least partial answers with mathematical modeling. For instance, one may ask how it is possible that, at the same time, isolated languages, such as Picard in Northern France or Quechua in Peru, have managed to survive, while other languages in similar situations have become extinct; or why Breton disappeared first in towns, whereas other languages disappeared first in rural areas. The linguistic environment is diversified by different 'ecological parameters', which are the main focus in this chapter; these parameters find respective counterparts in quantitative models, such as, the population size, the topology of the social network, the relative influence of prestige versus loyalty, the underlying physical or political geography, the population geography, and the economic environment (Patriarca et al., 2012). The quantitative representation of such ecological factors in mathematical models is usually simplified, but such clear-cut situations can help to fill the gap between the wide diversity of real situations observed by sociolinguists and a theoretical description. These aspects are addressed in the next sections by discussing a selection of simple models of language competition.

However, from a future perspective, the frontier of language dynamics is the modeling of language shift and language maintenance in the complex linguistic environment of global multilingualism. This is especially true in view of social applications that need to take into account various linguistic aspects, characterizing native as well as multilingual societies. In this respect, in planning policies defending and supporting minority languages, there is an interest in the smaller space scale and shorter time scales of real-life situations. This could be considered another field of language dynamics (besides language evolution and language competition), namely *language use*—some models of language use are briefly described in this chapter. In order to give an idea of the complexity of the situations to be described (and of the corresponding databases to be analyzed and understood in terms of language dynamics), we close this chapter with an example of a linguistic database taken from the study case of Basque—this is intended as an informative window on the types of data provided by linguists and as a challenge for the future work of mathematical modelers.

7.2 Two-State Models: The Abrams and Strogatz Model

In two-state models of language dynamics, there are two groups of monolingual speakers, a group speaking language X and the other speaking language Y. In the macroscopic version of two-state models, the system is described through population sizes $x(t)$ and $y(t)$ of the two groups at time t; in the individual-based version of two-state models with N individuals, each individual i ($i = 1, \ldots, N$) is either in the state X or in the state Y at time t. During the interaction, some individuals can adopt the other language, thus changing the values of the population sizes x and y, according to the scheme in Figure 7.1. A two-state model, as that represented in Figure 7.1 and with a constant total population evolves with time according to the equations

$$\begin{aligned}\dot{x} &= P(Y|X)\,y - P(X|Y)\,x, \\ \dot{y} &= -P(Y|X)\,y + P(X|Y)\,x,\end{aligned} \qquad (7.1)$$

where $\dot{x} = dx(t)/dt$ and $\dot{y} = dy(t)/dt$ are the time derivatives of $x(t)$ and $y(t)$ and the quantities $P(i|j)$ are functions of x and y representing the transition rates per individual from language i to language j (i, j = X,Y). The conservation of the total population, following from the fact that $\dot{x} + \dot{y} = 0$, allows one to rescale the population sizes x and y in a simple way so that they become the *fractions* of speaker of language X and Y, respectively, $x(t) + y(t) = 1$ at any time t, with $x, y \in (0, 1)$. The system in Eqs. (7.1) can now be reformulated through a single equation, for example, for the variable x,

$$\dot{x} = P(Y|X)(1 - x) - P(X|Y)x, \qquad (7.2)$$

where it is understood that in the transition probabilities P, which depend on x and y, the variable y is to be replaced by $(1 - x)$.

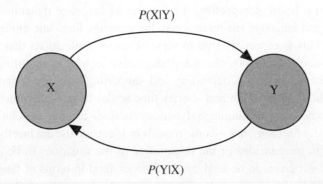

Figure 7.1 General scheme of a two-state language competition model describing two interacting languages X and Y. The quantities P represent transition rates at which an individual undergoes a language shift. In the AS (Abrams and Strogatz) model, $P(Y|X) = sx^a$ and $P(X|Y) = (1-s)(1-x)^a$ (see text for details).

Notice that each term in Eqs. (7.1), is proportional to the size of the 'recruited' population, since the total transition rate of a change is proportional to the size of the population of the agent undergoing the change of state. At the same time, the transition rate per individual is completely determined—together with the specific dynamical features of the model—by the transition rates $P(X|Y)$ and $P(Y, X)$.

The best known two-state model is the model of Abrams and Strogatz (2003) that had a relevant role in triggering the interest of a wide interdisciplinary community toward language dynamics modeling. It is a clear and simple macroscopic model in which the total population is conserved. The Abrams–Strogatz (AS) model assumes that $P(Y|X) \propto x^a$, where a is a coefficient; in this case, we refer to X as the 'attracting population'. The other rate is $P(X|Y) \propto y^a$, for which the attracting population is Y. The coefficient $a > 0$, called 'volatility parameter' (Castelló et al. [2006]), measures how easily an individual can change state from X to Y or vice versa. In statistical terms, the quantity $xP(X|Y)$ measures the total X→Y transition rate following the interactions between X and Y speakers (and analogously for the Y→X transition). To make an explicit example, in a homogeneous system where the interactions are determined by encounters between individuals, the encounter probability is proportional to the recruited population x and to the attracting population y, corresponding to a volatility $a = 1$, that is, the total transition rate is $xP(X|Y) \propto xy$, as in chemical kinetics (van Kampen, 1992). In fact, the explicit AS dynamical equations are similar to the reaction kinetics equation of two chemical species X and Y transforming into each other with rate coefficients $s_i > 0$ ($i = $ X,Y) and stoichiometric coefficients (or effective orders) given by a and 1,

$$\dot{x} = s_X x^a y - s_Y y^a x,$$
$$\dot{y} = -s_X x^a y + s_Y y^a x. \tag{7.3}$$

In a linguistic framework, s_X and s_Y are interpreted as a measure of the prestiges of the languages, for example due to their degree of official recognition; while the volatility parameter a ($a > 0$) measures how easily an individual can adopt another language (the 'volatility' is the corresponding property: a high volatility corresponds to a small value of the volatility parameter and vice versa).

Adopting a suitable time unit, the rate constants can be rescaled so that one can introduce the parameter $s = s_X/(s_X + s_Y)$, with $0 < s < 1$. Using the condition $x + y = 1$ and the parameter s, the AS model can be described through the single differential equation

$$\dot{x} = s x^a (1 - x) - (1 - s)(1 - x)^a x. \tag{7.4}$$

The properties of the motion defined by Eqs. (7.4) can be determined through stability analysis (see Strogatz [1994] for an introduction). The result is that the AS model system has both stable and unstable equilibrium points. Systematic studies of the AS model in different regimes of volatility have demonstrated the relevance of the volatility parameter a in determining the final state of the system (Castelló et al., 2006, 2011; Vazquez et al., 2010):

- *Low volatility.* In the regime of low volatility, that is, for values of the volatility parameters $a > 1$, there are two stable equilibrium points at $(x, y) = (0, 1)$ and $(x, y) = (1, 0)$, characterized by the presence of a single language and the extinction of the other one; and there is an equilibrium point at (x^*, y^*), with coordinate ratio $x^*/y^* = (1/s - 1)^{1/(a-1)}$, where the two languages coexist; this point is unstable for $a > 1$. In general, the value $a > 1$ leads to the stability only of the equilibrium points $(1,0)$ and $(0,1)$, implying the disappearance of one language. This situation corresponds to the results of the analysis carried out by Abrams and Strogatz (2003), with the goal of fitting data of Quechua (competing with Spanish); Welsh and Gaelic (competing with English). The fits provide values $a > 1$ for those cases. Abrams and Strogatz (2003) suggested a value $a \approx 1.3$, but both these values and the existence itself of a universal value has been questioned by more recent analyses (see e.g., Isern and Fort [2014] and Sutantawibul et al. [2018]). The actual problem of the volatility parameter behind any estimates of its numerical value is the lack of a reliable statistical, social, or linguistic interpretation.
- *High volatility.* The regime of high volatility is characterized by small values of the volatility parameter, $0 < a < 1$. The main difference with respect to the case of low

volatility at a microscopic level is that transitions are much more probable in this regime. This has a drastic consequence on the system dynamics and relaxation to equilibrium: the stable and unstable character of the equilibrium points are reversed, that is, the X–Y mixed state defined by the point (x^*, y^*) is now stable while the points $(0,1)$ and $(1,0)$ become unstable equilibrium points. The system can now live in a mixed X–Y state.

- *Neutral volatility.* Neutral volatility corresponds, by definition, to the critical volatility parameter separating the low from the high volatility regimes and is defined by the value $a_{\text{crit}} = 1$. In this case, Eqs. (7.3) simplify, reducing to (for $s_X > s_Y$, setting $s' = s_X - s_Y > 0$)

$$\dot{x} = +s' x y,$$
$$\dot{y} = -s' x y, \qquad (7.5)$$

in which only Y→X transitions are allowed: this means that the system relaxes toward the only stable equilibrium point $(x_1, y_1) = (1, 0)$, where only language X survives, while the other equilibrium point $(x_2, y_2) = (0, 1)$ is unstable, since any arbitrary small variation of population $x \to x' > 0$ would cause a transition toward point (x_1, y_1) and the disappearance of language Y (since $s' > 0$). Notice that only for $a \neq 1$, Eqs. (7.3) describe a system with transitions in both directions X↔Y, which represent two languages X and Y that are at the same time preys and predators (Murray, 2002). By eliminating the variable $y = 1 - x$ from Eqs. (7.5), one obtains the reduced equation

$$\dot{x} = s' x (1 - x), \qquad (7.6)$$

that is, as the logistic equation. This shows that the logistic equation can describe competition processes between two conventions in a population-conserving system, between a state X, in which the innovation is, for example, adopted, and the complementary state Y, in which it is not.

The dynamics of the AS system is depicted in Figure 7.2, representing the *phase portrait* corresponding to Eqs. 7.3 for the case $a > 0$. The two-dimensional xy plane is used here for convenience, in view of later comparisons; the one-dimensional x-axis is otherwise sufficient to study the stability of Eq. (7.4). In fact, the effective motion of the representative point is one-dimensional and, within the xy-plane, takes place along the line $x + y = 1$, defined by the constraint on the conservation of the total population.

When the representative point $x(t), y(t)$ in the xy-plane reaches a *stable equilibrium point*, that is, the point $(0,1)$ or $(1,0)$ in Figure 7.2 (for $a > 1$), it will stay there forever and, if perturbed (or moved close enough), it will go back to this stable point, as represented by the arrows. However, if the system is located at the unstable equilibrium point (x^*, y^*), it will stay there only if unperturbed; if an arbitrarily small perturbation is applied, which displaces the representative point to an arbitrarily small distance from (x^*, y^*), it will get further and further from the initial point, moving eventually toward one of the stable equilibrium points.

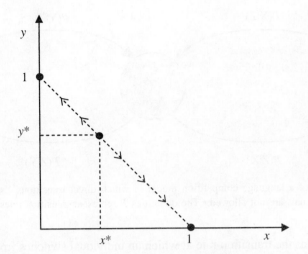

Figure 7.2 The dynamics of the AS model is shown in the two-dimensional xy space for later comparison: x and y are the population sizes of the X and Y group, the dots are the three equilibrium points. The representative point $(x(t), y(t))$ can move in time along the segment joining the $(0,1)$ to the $(1,0)$ points, where the total population is conserved, $x + y = 1$, $x, y \in (0, 1)$. Arrows represent the direction of the motion of the representative point for $a > 1$ (arrows are reversed for $a < 1$), showing that the equilibrium point (x^*, y^*) is unstable (stable) for $a > 1$ ($a < 1$), while the equilibrium points $(0,1)$ and $(1,0)$ are stable (unstable).

The initial conditions determine the final equilibrium point. For $a > 0$, there is only local (but not global) stability (Strogatz, 1994): for initial conditions $(x_0, y_0) = (x_0, 1 - x_0)$ with $x_0 < x^*$, the system evolves toward the equilibrium point $(0, 1)$, where only language Y survives; while starting from initial conditions with a large enough X fraction x_0, that is, $x_0 > x^*$, the system will evolve toward the equilibrium point $(1,0)$, where only language X survives.

The AS model has also been studied in its individual-based version, in regular lattices as well as in complex networks (see Stauffer et al. [2007]; Castelló et al. [2006]; Vazquez et al. [2010]; Castelló et al. [2011]), revealing very interesting dynamics corresponding to the different volatility regimes.

7.3 Models with Bilinguals

Two-state models can be extended to three-state models in order to include bilinguals. In such models, there are three populations: monolinguals X, monolinguals Y, and bilinguals, denoted by Z, associated with the population sizes x, y, and z, respectively, so that, technically, we now have a three-state model. In this section, the focus is on a special category of models, schematized in Figure 7.3, in which direct X↔Y transitions are absent: a monolingual (i.e., an individual in a state X or Y) can become a bilingual (switch state to Z) and vice versa but cannot turn directly into a monolingual of the other language. In order that a language shift takes place, a speaker has first to become a bilingual and then, eventually, become a monolingual of the other language.

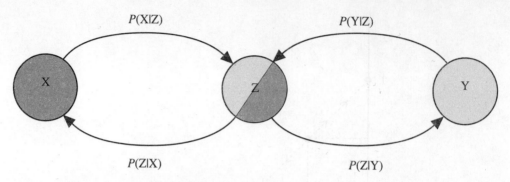

Figure 7.3 Scheme of a language competition model in which direct transitions X↔Y between different monolinguals are not allowed. The quantities P represent transition rates (see text for details).

Defining $P(i|j)$ as the transition rate at which an individual switches from state i to state j, the equation of the three-state model can be written as

$$\frac{dx}{dt} = P(Z|X)z - P(X|Z)x,$$
$$\frac{dy}{dt} = P(Z|Y)z - P(Y|Z)y. \tag{7.7}$$

There is actually a third equation for the fraction z of bilinguals, which is not written above, here; due to population conservation, one has the condition that $\dot{z} = -\dot{x} - \dot{y}$.

7.3.1 The Minett and Wang model

The Minett and Wang (MW) model is an extension of the AS model that also includes bilinguals (Wang and Minett, 2005; Minett and Wang, 2008). An outcome of the model was the proposal of a strategy for preserving endangered languages; the proposal consisted of changing the prestige of a language as soon as the density of speakers falls below a given value (intervention threshold) at which the language is to be considered in danger. This language policy can lead to a coexistence scenario with two languages.

The MW model describes both horizontal and vertical transmission of a language; to this aim, it contains various free parameters, including prestige, volatility, transition rates, and a mortality rate, although these parameters appear merged into a smaller number of effective parameters. The MW model is a three-state model of the type mentioned earlier, with three types of speakers: X monolinguals, Y monolinguals, and bilinguals, Z, and in which only transitions X↔Z and Y↔Z occur, while transitions X↔Y are not allowed (Figure 7.3). As in the AS model, it is assumed that the total population is conserved, $x(t) + y(t) + z(t) = 1$, at any time t; then, only two variables, for example, x and y, are sufficient to describe the system. At variance with the two-state population-conserving models, the representative point $(x(t), y(t))$ in the xy space can now explore the whole triangle defined by $x, y \geq 0$ and $x + y \leq 1$. The model can be defined by the two equations for the variables x and y,

$$\frac{dx}{dt} = k_{ZX}xz - k_{XZ}yx,$$

$$\frac{dy}{dt} = k_{ZY}yz - k_{YZ}xy. \tag{7.8}$$

Here, the parameters k_{ZX}, k_{XZ}, k_{ZY}, k_{YZ} represent effective transition rates and the terms describe how x and y vary due to transitions between bilingual and monolingual communities. Notice that the rate at which bilinguals become monolinguals X (or Y) is proportional to x (y); hence, this model represents a generalization of the AS model in the particular case of neutral volatility, $a = 1$. Eqs. (7.8) are equivalent to the general Eqs. (7.7), if the transition rates are given by

$$P(X|Z) = k_{XZ}y, \qquad P(Z|Y) = k_{ZX}y,$$

$$P(Y|Z) = k_{YZ}x, \qquad P(Z|X) = k_{ZX}x. \tag{7.9}$$

Equations (7.9) show that the transition rates between monolinguals and bilinguals are determined solely by the population sizes of *monolinguals* x and y.

By replacing $z = 1 - x - y$ in Eqs. (7.8), one obtains the closed equations

$$\frac{dx}{dt} = x[k_{ZX}(1 - x - y) - k_{XZ}y],$$

$$\frac{dy}{dt} = y[k_{ZY}(1 - x - y) - k_{YZ}x]. \tag{7.10}$$

The study of the dynamics of the model reveals two stable equilibrium points, $(x_1, y_1, z_1) = (1, 0, 0)$, where all people are monolinguals in X; and the complementary equilibrium point $(x_2, y_2, z_2) = (0, 1, 0)$, where all people are Y speakers. Furthermore, there are two unstable equilibrium points: $(x_3, y_3, z_3) = (0, 0, 1)$, corresponding to a system where all individuals are bilinguals (represented by the origin of the xy-plane, $x = y = 0$ and $z = 1$), and a fourth equilibrium point containing speakers of all three types X, Y, and Z, with coordinates $(x_4, y_4, z_4) = (k_{XZ}k_{ZY}/\Sigma, k_{ZX}k_{YZ}/\Sigma, k_{XZ}k_{YZ}/\Sigma)$, where $\Sigma = k_{XZ}k_{ZY} + k_{ZX}k_{YZ} + k_{XZ}k_{YZ}$. Somewhat surprisingly, the inclusion of bilinguals in this new model cannot prevent the relaxation toward monolingual situations, accompanied by the disappearance of one language, as in the AS model. Figure 7.4 (upper panel) shows the phase portrait with two trajectories starting with initial conditions $(x(0), y(0), z(0)) = (0.6, 0.4, 0)$ and $(x(0), y(0), z(0)) = (0.55, 0.45, 0)$, i.e., initiallly there are no bilinguals, and ending at the equilibrium point (x_1, y_1, z_1) and (x_2, y_2, z_2), respectively.

It can be noticed that Figure 7.4 (lower panel), representing the time evolution of the population sizes $x(t)$, $y(t)$, and $z(t)$ for the trajectory ending at (x_1, y_1, z_1), shows that the system is in a state with a majority of bilinguals over a long time span (notice the logarithmic scale of the time axis). The bilingual population eventually decreases, but only when the monolinguals group X has already disappeared. The latter feature can be understood from Eqs. (7.8) by

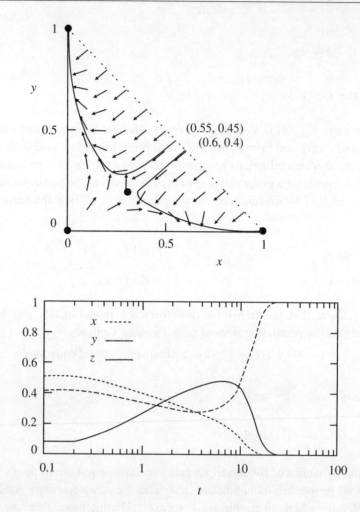

Figure 7.4 MW model (Eqs. [7.8] or [7.10]) with parameters $k_{ZX} = 0.4$, $k_{ZY} = 0.5$, $k_{YZ} = 0.8$, $k_{XZ} = 1$. Upper panel: Velocity field (arrows) in the accessible triangle $x, y \geq 0$, $x + y \leq 1$ and two trajectories (solid lines), one leading to the equilibrium point $(0,1)$ (only language Y), the other to $(1,0)$ (only language X). Since $z(0) = 0$, the trajectories start from the $x + y = 1$ line at the $x(0)$, $y(0)$ values indicated. Lower panel: Population sizes versus time for $x(0) = 0.55$, $y(0) = 0.45$, $z(0) = 0$. Reprinted from Heinsalu et al. (2014) with the permission of World Scientific.

noting that for $x \to 0$, one obtains $dz/dt \approx -dy/dt \approx -k_{ZY}yz$, implying a fast decrease of $z(t)$. The existence of such a long time span, in which the system is mostly populated by bilinguals, questions the validity of one of the basic hypotheses underlying the MW model, namely, that the transition rates do *not* depend on $z(t)$ but solely on the monolingual population sizes $x(t)$ and $y(t)$ or that, in other words, only monolinguals can play the role of attracting populations. This is a delicate question and will be discussed further in the following sections.

7.3.2 Connecting language and opinion dynamics: The AB model

The 'AB model' is a language competition model introduced by Castelló et al. (2006) that was inspired by the AS model of language competition[1] and constructed on the base of the voter model, a well-known opinion dynamics model (Holley and Liggett, 1975; Vazquez et al., 2010; Fernández-Gracia et al., 2014). The AB model was formulated as a microscopic individual-based model on a general network and, as such, presents a rich dynamics (see Castelló et al. [2007]; Vazquez et al. [2010]; Castelló et al. [2011]) which is beyond the goal of this chapter. In this section, we recall the mean field limit of the AB model (in a fully connected network) and compare it with other models of bilinguals.

The mean field equations for the population sizes, assuming a neutral volatility and a conserved total population ($z = 1 - x - y$), are

$$\frac{dx}{dt} = s(1-y)z - (1-s)xy,$$
$$\frac{dy}{dt} = (1-s)(1-x)z - sxy. \qquad (7.11)$$

These equations can be rewritten as Eqs. (7.7) by setting

$$P(X|Z) = (1-s)y, \qquad P(Z|Y) = (1-s)(1-x),$$
$$P(Y|Z) = sx, \qquad P(Z|X) = s(1-y). \qquad (7.12)$$

Here, the monolingual→bilingual transition rates $P(X|Z)$ and $P(Y|Z)$ are equal to those of the MW model (Eqs. 7.9), if one sets $k_{XZ} = (1-s)$ and $k_{YZ} = s$, while the bilingual→monolingual rates are different, being $P(Z|X) \propto (1-y) = x + z$ and $P(Z|Y) \propto (1-x) = y + z$, in which the attracting populations are the *total populations that can speak the language* X and Y, respectively, that is, monolinguals (X or Y) *and* bilinguals.

It can be shown that the model has three fixed points, the first two points coinciding with the monolingual states of the MW model and a third unstable equilibrium point, where all the three populations are present. Thus, the main conclusions of the AS and MW models, concerning the disappearance of one language, are confirmed in the AB model; in addition, it was shown that the presence of bilinguals generally reduces the scenario of language coexistence with respect to the MW model (Castelló et al., 2006), by accelerating the disappearance of the minority language. The reason lies in the fact that the bilingual→monolingual transition rates grow with the *whole* population that speaks the new (adopted) language, including bilinguals, and not only with that of the monolinguals. At a microscopic level, this means that a Z→Y transition can be triggered not only by a Y–Z encounter between a monolingual Y and a bilingual Z, but also by a Z–Z encounter between two bilinguals (who use language Y to communicate with each other).

---

[1] A and B were the labels used for indicating the two languages or opinions, here referred to as X and Y.

The AB model has also been studied on complex networks and provides a nice example of the deep influence that a complex underlying topology can have on the system dynamics. The case of the AB model on a complex network is particularly interesting, since, as discussed earlier, the mean field limit of the model in a fully connected network converges fast to a state populated by one monolingual group. Castelló et al. (2007) studied the voter model and the AB model on several networks. In most of the cases, they found a similar relaxation toward a monolingual state. However, when considering a social model of a network with communities, characterized by a large heterogeneity at the mesoscale level, they found the surprising result that the presence of bilinguals can slow down the relaxation of the system so effectively as to produce a power law relaxation time distribution; this means in practice that there is actually no (relaxation) time scale and that, at any arbitrary long observation time, both monolinguals and bilinguals are present (Toivonen et al., 2009; Castelló et al., 2007). No such effect was found on any type of network for the voter model, that is, a two-state model with excluding options.

7.3.3 The role of bilinguals

In this section, some generalizations of the MW model are discussed. They reveal the possible relevance of bilinguals in language competition. The MW model assumes that only monolinguals can play the role of the 'attracting' populations, meaning that only monolinguals can induce a state change in a speaker—either bilingual or monolingual. This is reflected by the dependence of the corresponding transition rates on the monolinguals' population size only. For example, in the MW model, the Z→Y transition rate $P(Z|Y) \propto y$ (see Eqs. [7.9]), meaning that the presence of Y speakers induces bilinguals to use only language Y and they finally become Y monolinguals. Even if the bilingual state Z represents a necessary step along a complete X↔Y language shift, bilinguals do not influence the transition rates. This hypothesis can be questioned—as already mentioned at the end of Section 7.3.1—in relation to the fact that bilinguals may become at some point and remain for a long time the majority of the population (see Figure 7.4).

One can expect that the introduction of a direct dependence of the monolingual → bilingual transition rates on the bilingual population should be able to trigger some new type of X→Z transitions. In the model introduced by Heinsalu et al. (2014), bilinguals are considered to be representatives of both the languages they speak, so that they enter into the dynamics on the same footing as monolinguals; correspondingly, the monolingual → bilingual transitions are now associated with the following rates,

$$P(X|Z) = k_{XZ}(y + z), \qquad (7.13)$$

$$P(Y|Z) = k_{YZ}(x + z). \qquad (7.14)$$

Considering the first equation, for example, one has the condition that the X→Z transition is not only determined by the presence of or encounters with a Y monolingual but that a bilingual individual Z will have the same effect. The stability analysis of the three-state model defined by the transition rates, Eq. (7.14), provides four equilibrium points (Heinsalu et al., 2014):

$(x_1, y_1, z_1) = (1, 0, 0)$, only X monolinguals

$(x_2, y_2, z_2) = (0, 1, 0)$, only Y monolinguals

$(x_3, y_3, z_3) = (0, 0, 1)$, only bilinguals

$(x_4, y_4, z_4) = (k_{XZ}(k_{ZY} - k_{YZ})/\sigma, k_{YZ}(k_{ZX} - k_{XZ})/\sigma, k_{XZ}k_{YZ}/\sigma)$,

where $\sigma = k_{XZ}k_{ZY} + k_{ZX}k_{YZ} - k_{XZ}k_{YZ}$.

The first and second equilibrium point (already predicted by the AS model), corresponding to an X or Y monolingual system, respectively, can be locally or globally stable, or unstable, depending on the values of the transition rates. The third equilibrium point, representing a system with only bilinguals, can also become globally stable—at variance with the MW model—when $k_{XZ} > k_{ZX}$ and $k_{YZ} > k_{ZY}$, that is, when the monolingual→bilingual transition rates are larger than those for the opposite transitions. The last equilibrium point, corresponding to a mixed system where all groups are represented, exists only if $k_{ZY} > k_{YZ}$ and $k_{ZX} > k_{XZ}$ and is unstable for all parameter values. The phase portraits and the trajectories of the representative point $(x(t), y(t))$ in the xy-space for different sets of parameters four different scenarios corresponding to these are represented in Figure 7.5. Figure 7.6 illustrates the time dependence of the population sizes for the trajectory that brings the system to the equilibrium state of bilinguals only (Figure 7.5, panel [a]).

A more realistic scenario for the transitions from mono- to bilinguals, somewhere between that defined by the transition rates in Eq. (7.14) and the MW model, can be described using the transition rates

$$P(X|Z) = k_{XZ}(y + \alpha z), \quad P(Y|Z) = k_{YZ}(x + \beta z), \tag{7.15}$$

where the parameters $\alpha, \beta \in [0, 1]$ modulate the relative importance of bilinguals with respect to monolinguals, as representatives of language Y and X, respectively (this model reduces to the model considered earlier for $\alpha = \beta = 1$ and to the MW model for $\alpha = \beta = 0$). The model has four equilibrium points (Heinsalu et al., 2014), the first three coinciding with those obtained earlier for $\alpha = \beta = 1$, and a fourth equilibrium point,

$$(x_4, y_4, z_4) = \left(\frac{k_{XZ}(k_{ZY} - \beta k_{YZ})}{\sigma_{\alpha\beta}}, \frac{k_{YZ}(k_{ZX} - \alpha k_{XZ})}{\sigma_{\alpha\beta}}, \frac{k_{XZ}k_{YZ}}{\sigma_{\alpha\beta}} \right),$$

where $\sigma_{\alpha\beta} = k_{XZ}(k_{ZY} - \beta k_{YZ}) + k_{YZ}[k_{ZX} + k_{XZ}(1 - \alpha)]$,

which exists only for $k_{ZY} > \beta k_{YZ}$ and $k_{ZX} > \alpha k_{XZ}$ and is always unstable. The stability properties of the first three equilibrium points are determined by special values of α and β,

$$\alpha_{\text{cr}} = k_{ZX}/k_{XZ}, \quad \beta_{\text{cr}} = k_{ZY}/k_{YZ}.$$

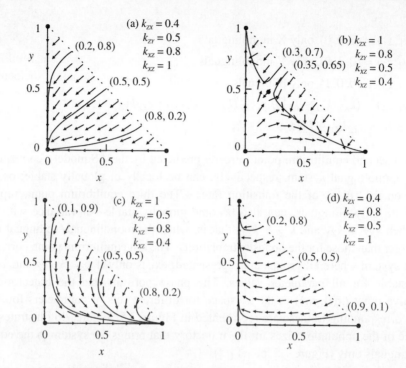

Figure 7.5 Generalized MW model with rates given by Eqs. (7.14): Phase portraits and sample trajectories for the four different scenarios. Panel (a): $k_{XZ} > k_{ZX}$, $k_{YZ} > k_{ZY}$, the system converges toward a globally stable bilingual community; Panel (b): $k_{ZX} > k_{XZ}$, $k_{ZY} > k_{YZ}$, monolinguals X or Y societies are locally stable states; Panel (c): $k_{ZX} > k_{XZ}$, $k_{YZ} > k_{ZY}$, the X monolingual community is a globally stable equilibrium state; Panel (d): $k_{ZY} > k_{YZ}$ and $k_{XZ} > k_{ZX}$, the Y monolingual community is a globally stable equilibrium state. Reprinted from Heinsalu et al. (2014) with permission from World Scientific.

Depending on these critical values, the following scenarios are possible.

- $\alpha > \alpha_{cr}$ and $\beta > \beta_{cr}$: The final state is a globally stable equilibrium point that is solely bilingual (see Panel (a) in Figure 7.7).
- $\alpha < \alpha_{cr}$ and $\beta < \beta_{cr}$: Either language X or Y will survive (as in the MW model), depending on the initial values $x(0)$ and $y(0)$ (see Panel (b) in Figure 7.7).
- $\alpha < \alpha_{cr}$ and $\beta > \beta_{cr}$: An X monolingual group is globally stable (independently of the initial conditions) (see Panel (c) in Figure 7.7).
- $\alpha > \alpha_{cr}$ and $\beta < \beta_{cr}$: A Y monolingual group becomes a globally stable system (independently of the initial conditions) (see Panel (d) in Figure 7.7).

For additional details see Heinsalu et al. (2014). Though it is difficult to determine the exact dependence of α and β on the various external factors in a real situation, a bilingualism policy can be expected to increase the values of α and β. A larger value of α and β, in turn, may have an impact on language survival, as shown by the existence of critical values of α and β that ensure the survival of the respective language. If both the parameters are larger than their critical thresholds, both languages will certainly survive, a scenario which is more interesting

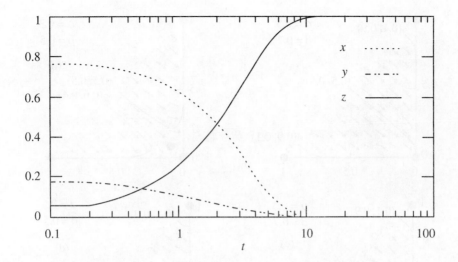

Figure 7.6 Time evolution of the population sizes $x(t), y(t), z(t)$ for the trajectory in Panel (a) of Figure 7.5 with initial condition $x(0) = 0.8$, $y(0) = 0.2$, $z(0) = 0$, ending in a stable bilingual community ($x = y = 0$). Notice the logarithmic scale of the time axis. Reprinted from Heinsalu et al. (2014) with permission from World Scientific.

and beyond the one of a single winner between two competitors. Specific linguistic policies aimed at protecting and revitalizing minority languages can try to change the values of α and β 'on the fly', to change the dynamics of a system relaxing toward a monolingual state into a new one that makes the system evolve toward a bilingual solution.

A possible objection to the conclusions drawn from the discussion of this model is that, usually, no purely bilingual community is observed in real situations. An answer to this objection can be based on two relevant points.

The first point is that the hypothesis that bilingualism leads to monolingualism, as predicted by the Minett and Wang model (and advocated by Aracil [1982]) is not confirmed by observations. We have mentioned the general trend toward global multilingualism; in such a multilingual reality, there are some languages that are spoken by the majority, side by side with some other languages that are spoken in specific environments. For example, in the interactions between Castillan and Catalan language or in some areas of the Gipuzkoa in the Basque Autonomous Community, the sociolinguistic paradigm is committed to an integrative approach of social networks (Palau, 2002). Bilingualism is usually observed if there is a competition between the official language and a protected minority language. The situation observed is usually dynamic and complex, even when it does not show clear-cut examples of bilingualism or multilingualism and the stable bilateral bilingualism of theoretical models is still an ideal situation: for example, Quebec's bilingualism is still far from stable and is quite conflictual; Switzerland is in some respects more a confederation of bilingualisms than a stable multilingual community; Belgium's bilingualism may be decreasing, following the fact that Eastern Belgium is not economically dominated by Western Belgium anymore.

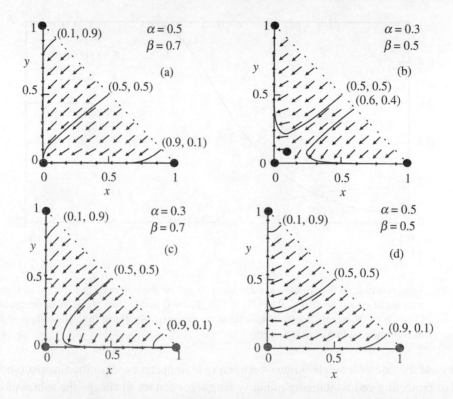

Figure 7.7 Phase portrait and some trajectories for the generalized model with parameters α and β described by Eqs. (7.15) for the four possible scenarios. Here the k-parameters, equal to those of Figure 7.6, and the critical values, $\alpha_{cr} = 0.4$ and $\beta_{cr} = 0.625$, are constant through all the panels. Panel (a): $\alpha > \alpha_{cr}$, $\beta > \beta_{cr}$, the final state of the system is always a bilingual community. Panel (b): $\alpha < \alpha_{cr}$, $\beta < \beta_{cr}$, either language X or Y survives, depending on the initial conditions $x(0)$ and $y(0)$. Panel (c): $\alpha < \alpha_{cr}$, $\beta > \beta_{cr}$, the final state is always an X-monolingual group. Panel (d): $\alpha > \alpha_{cr}$, $\beta < \beta_{cr}$, the final state is always an Y-monolingual group. Reprinted from Heinsalu et al. (2014) with permission of World Scientific.

The second point is technical in nature but is closely related to the importance of ecological factors in determining the final linguistic scenario. The simple models discussed so far describe *homogeneous* societies and usually have homogeneous equilibrium solutions, representing prototypical situations with either the community of monolinguals of one language or the bilingual community only. More realistic inhomogeneous equilibrium solutions with fragmented states are encountered when introducing heterogeneous elements (ecological factors) at some level of the parameter space of the model: examples are geographical barriers (Patriarca and Leppänen, 2004; Patriarca and Heinsalu, 2009), a language prestige that is space-modulated (Kandler, 2009), or heterogeneous population dynamics (Baggs and Freedman, 1990, 1993; Pinasco and Romanelli, 2006). Some example of models with heterogeneous ecological parameters are considered in the following sections. In addition, it is to be noticed that there are also simple models of language dynamics that predict fragmented scenarios and persistence as a mere consequence of the intrinsic properties of their language dynamics (without additional factors such as an underlying population dynamics process), as in the AS model with a high volatility ($a < 1$) or in the naming game model with a low acceptance probability $\beta < 1/3$, considered in Section 7.3.4.

7.3.4 The naming game as a competition model

The naming game (NG) was discussed in Section 6.1.2 in relation to individual-based models of language evolution since in its original formulation, it represents the evolution of a language model seen as a set of couplings between concepts and names (Loreto and Steels, 2007). As mentioned earlier, the NG model—actually any semiotic model—can be reformulated as a competition model using the population sizes of each language, rather than the elements of the lexical matrices, as main variables (Castellano et al., 2009). When considering the interaction between a single concept and many names, as in the NG, the structure of the competition process simplifies and in the particular case of the two-conventions NG, a very close and illuminating analogy with three-state competition models appears. An agent who knows and uses only word A (or B) or both words A and B in a semiotic model corresponds in a three-state competition model to—and will be referred to in the following as—a monolingual X (or Y) or bilingual (Z), respectively. In order to bring out such an analogy, it is sufficient to consider the mean field limit of the individual-based update rules of the NG. This is illustrated in Castelló et al. (2009), where also a generalized form of the NG model was introduced, in which an agent makes the usual agreement with another agent only a fraction β of times, while skipping it (and remaining in the original state) in the remaining fraction of times $1 - \beta$; the basic NG model is retrieved when $\beta = 1$. The introduction of the acceptance (or 'trust') parameter β has major consequences on the dynamics of the NG model: the appearance of an order–disorder phase transition taking place when the acceptance parameter crosses the critical value $\beta^* = 1/3$. For $\beta < \beta*$, there is no convergence toward a shared word and no equilibrium is ever reached, with the system continuing to live between different states fragmented into three groups of X, Y, and Z individuals.

The mean field equations, obtained by considering all the possible inter-agent encounters and the corresponding outcomes, are (Castelló et al., 2009)

$$\frac{dx}{dt} = -xy + \beta z^2 + \frac{3\beta - 1}{2}xz \equiv -\left(y + \frac{z}{2}\right)x + \beta\left(z + \frac{3}{2}x\right)z,$$

$$\frac{dy}{dt} = -xy + \beta z^2 + \frac{3\beta - 1}{2}yz \equiv -\left(x + \frac{z}{2}\right)y + \beta\left(z + \frac{3}{2}y\right)z. \tag{7.16}$$

Here, it is understood that the bilingual population size is $z = 1-x-y$ due to the total population being constant. These equations become identical to the general Eqs. (7.7) if we identify the transition rates as follows:

- Monolingual→bilingual transitions: $P(X|Z) = (y + z/2)$ and $P(Y|Z) = (x + z/2)$ are enhanced (1) by the monolinguals of the other language (y and x respectively), as in the MW model, and (2) also by the bilinguals, as in the generalized model introduced in Section 7.3.3.

- Bilingual→monolingual transitions: $P(Z|X) = \beta(z + 3x/2)$ and $P(Z|Y) = \beta(z + 3y/2)$ depend (1) on the monolingual population size, x and y, respectively, and (2) also on the bilingual population size, as in the AB model.

Notice that the bilingual↔monolingual transition rates in both directions depend on the bilingual populations size, z. This double dependence is peculiar to the NG and is present also in the basic version, for $\beta = 1$. Despite the relative frequencies of the various types of encounters and interactions being fixed, the NG model considers a topology of interactions that is unique among the other language competition models: in particular, (a) a bilingual may become a monolingual by interacting with bilinguals (as in the AB model) and (b) a monolingual may become a bilingual by interacting with bilinguals (as in the model introduced in Section 7.3.3). However, no order–disorder transition is observed, neither in the basic NG model ($\beta = 1$), nor in the AB model, nor in the model described in Section 7.3.3, meaning that both the double dependence on z and the β modulation of the monolingual→bilingual transition are necessary elements for the existence of a fragmented state where all the linguistic groups, the two monolingual groups X and Y and the bilingual group Z, can coexist.

In spite of the fact that many transition constants are set equal to each other and the acceptance parameter β is the only free parameter of the model, by modulating β one can trigger the order–disorder transition and the appearance of a mixed-state equilibrium. It is to be expected that a generalized model with arbitrary transition rates, built using the NG model as base, may have a rich set of linguistic scenarios as the possible final equilibrium states. Thus, the NG model originally introduced in semiotic dynamics, has unexpectedly turned out to be a valuable language competition model, suited to study most different situations.

In fact, a simple generalization of the basic NG model, the ABC model, a three-language model put forward by Pucci et al. (2014), was applied to the study of the appearance of creole languages by Tria et al. (2015). The ABC model describes the interactions of three different languages A, B, and C, with the possibility of having bilinguals of different types with different combinations of languages, with the specific feature that a new language C can appear in the bilinguals' group. Although the appearance of a contact language is a topic more appropriately discussed within language evolution, technically, the model considers the two original languages A and B in contact as well as the emerging creole language C as fixed entities, describing an actual competition process. Historically (see Tria et al. [2015] for details), the background of the emergence of creole languages was a multilingual society made up of (a) Europeans, speaking only an European language, (b) Bozal (slaves), speaking a set of quite different (and relatively non-mutually intelligible) African languages, and (c) Mulattos (free Blacks), representing the main interface in the communications between Europeans and Bozal and assumed to speak initially the same European language. The system evolves with time following the same rules of the NG with acceptance parameter β; the probabilities of success are suitably modified so that they are adapted to the specific features of the communications between Europeans and Bozal and between different Bozal languages. The simple yet surprising result is that either the European language or creole eventually spreads across the Mulattos and the Bozal groups and that which language spreads basically depends only on the respective population fractions n_B and n_M of Bozal and Mulattos. The authors studied different historical locations and showed that they can be partitioned into two groups,

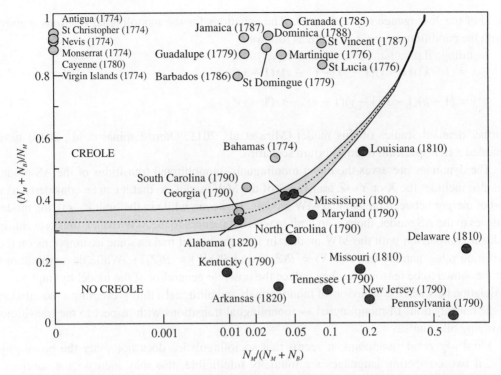

Figure 7.8 Locations where a creole language appeared (light circles) or did not appear historically (dark circles) in the plane of the variables $n_M = N_M/(N_B + N_M)$, the fraction of Mulattos, and the combined fraction of Mulattos and Bozal, $n_B + n_M = (N_M + N_B)/N = (N_M + NB)/(N_M + N_B + N_E)$, where M, B, and E refer to Mulattos, Bozal, and Europeans, respectively. The gray stripe dividing the two groups is predicted on the basis of the theoretical model for a suitable set of parameters. (From Tria et al. [2015]: Tria, F.; Servedio, V. D.; Mufwene, S. S. and Loreto, V., *Modeling the Emergence of Contact Languages*, PLOS ONE, Public Library of Science, 2015, 10, 1-11. doi:10.1371/journal.pone.0120771)

the group in which a creole language appeared or the group in which it did not appear, on the basis of the values of the population fractions n_B and n_M (see Figure 7.8).

7.3.5 Similarity between languages: The Mira and Paredes model

The model introduced by Mira and Paredes (2005) (see also Mira et al. [2011]; Otero-Espinar et al. [2013]; Nie et al. [2013]; Colucci et al. [2016] for more recent developments) studies the role of a possible similarity between the dominant and the dominated language. The model raises the more general question of the interaction between very similar, mutually intelligible languages, and provides a quantitative scheme for its description. Notice that this model is not described by the scheme in Figure 7.3, due to the existence of direct monlingual↔monolingual transitions, but is closely related to it. The model depends on the parameter $k \in (0, 1)$ that represents the fraction of direct X–Y transitions, with respect to X–Z and Y–Z transitions (thus quantifying the similarity between languages X and Y) and on the language status $s \equiv s_X \in$

(0, 1) of the X language ($s_Y = 1 - s_X$). The equations for the monolingual fractions x and y (with the condition $x + y + z = 1$) are as follows.

$$\dot{x} = (1 - k)s(1 - x)(1 - y)^a - (1 - s)x(1 - x)^a,$$

$$\dot{y} = (1 - k)(1 - s)(1 - y)(1 - x)^a - sx(1 - y)^a. \tag{7.17}$$

Further detailed studies of this model (Mira et al., 2011; Otero-Espinar et al., 2013) have revealed a rich spectrum of equilibrium scenarios.

The dynamics preserves the direct monlingual↔monolingual transitions of the AS model but also includes the X or Y↔Z transitions of the MW model, so that it can be considered as a sort of merger between the AS and MW models. However, while in the limit $k \to 0$, the model reduces to the AS model, there is no limit in which it reduces to the MW model, thus preventing a direct comparison with the MW model. In fact, the model makes some assumptions on the transition rates, namely that $P(Y|X) = P(Z|X)$ and $P(X|Y) = P(Z|Y)$. While the assumptions can be shown to be reasonable, they reduce the possible generality of the model by implying a correlation between the behaviors of monolinguals and bilinguals, thus preventing a modulation of the contributions of monolingual ↔ monolingual transitions with respect to the transitions involving bilinguals.

From a general standpoint, it seems that sociolinguistics does not study the possibility that, if two competing languages are mutually intelligible, this may increase the surviving probability of the weaker language; this could be because one can observe also the opposite situation. In a positive perspective, in conditions of additive bilingualism and political actions, which favor a minority language and promote it to an asset in social relationships and work, the existence of similarities favor the spreading of the minority language. This is also the hypothesis assumed by Mira and Paredes (2005), which is probably correct for the case of the Galician language, but is not true in general. There are many counter-examples. A partial counter-example is Galician itself, which is mutually intelligible with Spanish, when compared, for example, with Basque, which is not mutually intelligible with Spanish but is more successful after many years of cultural autonomy. More extreme examples typically show how a language Y, mutually intelligible with a language X, is penalized just because of its similarity with the high-status language X that makes it look as a low-status dialect of X. For example, this was the case in the Oïl languages (or Oïl dialects) of Northern France.

However, the fit carried out using only k and s as free parameters successfully describe the data about the speakers of Galician, the speakers of Castillian, and the corresponding bilingual group in Galicia (Mira and Paredes, 2005). In fact, whatever the complex situation underlying two similar languages, the model would be able to accomodate all the various possible scenarios, including those in which the similarity between languages can represent a disadvantage, through its two parameters k and s, which allow to change not only the rate but also the direction of the direct monolingual ↔ monolingual transitions.

7.3.6 Use of language

As discussed in Section 7.1, in many cases, the way languages are used, that is, how often and where a bilingual speaker uses language X instead of language Y, is a more relevant question than the questions concerning which language will eventually be the 'winner' in the competition process between language X and language Y. In fact, the latter question is not even well posed in many real situations. For example, in many regions, where a language policy supports a minority language, bilingualism is observed and usually the minority language is not under immediate threat. In these situations, one is interested in the close future of that system, that is, the time evolution of the system on a shorter time scale in order to correct or make language policy planning. This type of knowledge depends on the finer details of the underlying language dynamics, for example, how speakers select the language used, what are the factors determining their decisions, and how often speakers use one or the other language.

The *conversation model* of Iriberri and Uriarte (2012) (referred to as the IU model hereafter) is a definite step in the direction of the study of language use (see also Uriarte [2016]). The IU model is more focused on language use than on language shift and therefore, particularly suited to the analysis of bilingual environments. In fact, the motivation for the introduction of the IU model was to gain a more appropriate description of multilingual societies that other language competition models simply could not provide.

The model also offers an insight for all other models of language dynamics, since it investigates the microscopic origin of the various transition rates by analyzing the behavior of speakers at the level of pair-wise conversation. For example, in the generalized model of bilinguals considered in Section 7.3.3, in which the influence of bilinguals is weighted by the coefficients α and β, Eq. (7.15), the IU model suggests a microscopic meaning of those coefficients in terms of the fraction of times that a bilingual decides to use language X and Y, thus acting as the 'attracting' individual in enhancing the adoption of the language used. Therefore, the IU model poses the more general problem concerning the macroscopic description and the large-scale consequences of the choices and strategies used by single speakers in their conversation for selecting the language used.

In order to model conversations involving bilinguals, the IU model is formulated as a language conversation game (see Uriarte [2012] for details). The model presents a mixed-strategy Nash equilibrium that is evolutionarily stable, consisting of two strategies used by a bilingual, with suitable respective rates p_a and $p_b = 1 - p_a$:
(a) a strategy consisting of using the minority language Y always and
(b) a strategy in which the minority language Y is used only when one knows for certain that the other individual is a bilingual; otherwise, X is used.

Considering this language game at the level of population, it means that the bilingual population Z is divided into two sub-populations Z_a and Z_b, whose individuals use the corresponding strategies. Such a game-theoretical model leads to a mixed equilibrium that is far from the simple equilibrium monolingual scenario of other models. The final solution, depending on the values of the parameters, can represent very different levels of use of the minority language, from the lowest one, where the majority language is used almost always also by bilingual speakers, to the level, in which bilingual speakers use their original language most of the times.

We mention also the model of Carro et al. (2016) because it is focused, like the IU model, on the use of language (but it would be otherwise difficult to categorize it as a two- or three-state model). The Carro et al. model is an individual-based model with N agents, in which speakers represent the nodes of a socially motivated model of network and their communications represent the links between them. The role of the network topology is crucial for the properties of the model; the network is constructed through a mechanism of triadic closure, which produces a set of connected triangles.[2] The use of language between an agent i and an agent j ($i, j = 1, \ldots, N$) is a property of the i–j link (i.e., of the interaction, rather than that of the node/speaker) and is represented by a dichotomic variable S_{ij} that can assume either the value $S_{ij} = 0$ (for language X) or $S_{ij} = 1$ (for language Y), depending on the language used. Instead, the preference of a speaker i to speak a language is a property of the speaker and is described by another variable $x_i \in (0, 1)$, with the value $x_i = 0$ meaning a maximal preference for language X and the value $x_i = 1$ meaning a maximal preference for language Y. The network topology is fixed but the system undergoes a coupled evolution of node and link states, that is, of the variables x_i and S_{ij}, under the interplay of two trends of the speakers, who try to: (a) minimize the cognitive effort associated with using several languages (i.e., they tend to use the same language always); (b) use the preferred language (see Carro et al. [2016] for details).

An interesting result is a complex asymptotic state definitely different from the simple scenario of a plain language shift in which only one language survives. In particular, it is found that the probability that a language disappears decreases exponentially with the system size N and that the survival time scale of dynamical metastable states increases linearly with N. Dynamical coexistence of the two languages is the natural asymptotic state for large enough systems, even if eventually one of the two languages dominates, while the other remains confined in small groups of bilinguals with a strong preference for the minority language—that they use for internal communications—but who regularly use the dominant language for communications with individuals outside their group.

Another model based on the triadic closure, which makes a detailed analysis of the multi-linguistic landscape in Estonia, is introduced and studied in Karjus and Ehala (2018)

The interesting common fact about these models focused on language use is that when the details of the communication between individuals are described in greater detail, the disappearance of a language is replaced by a wide spectrum of possible scenarios in which both languages survive in a wide set of different possible proportions and conditions.

We also mention the work of Bakalis and Galani (2012), in which two models were used to study the Aromanian, an endangered language in Greece and analyze the data extracted from the official census of Greece. One model was a two-state model with populations X and Y (corresponding to Greek and Aromanian) based on an (exactly solvable) approximation of the AS model for small values of the population size of the minority population. The second

---

[2] The triadic closure is the property characterizing the tendency of an individual to interact and create connections with other individuals with whom they share common connections.

model was an inhomogeneous (or structured) three-state model with populations X, Y1, and Y2. The latter model is noteworthy since it manages to take into account a relevant side of the social structure within the minimal framework of a competition model (thus connecting in spirit to the other models discussed in this section) by dividing bilinguals into two sub-groups, which have or do not have the tendency to communicate with each other in Aromanian. The detailed analysis produces a good fit of data revealing a worrying trend for the future of the Aromanian language, which is shown to be already in a very critical state. The time evolution is characterized by a volatility parameter close to the values originally found by Abrams and Strogatz (2003) for other database, $a \approx 1.2$.

7.4 Models with Population Dynamics

In the competition models considered in the previous sections, it was assumed that the total population is conserved, that is, $x + y = 1$ in the two-state models and $x + y + z = 1$ in the three-state models. This may look as a drastic simplification on the actual time evolution of a linguistic group but is an acceptable approximation in some cases, for example, in the presence of homogeneous growth, when the population has already reached its equilibrium size. Many models, also in different fields, are based on a microscopic interpretation of the dynamics in terms of a younger generation that on average replaces an older one at each time iteration, without actually changing the total population size.

Similar considerations can be done not only for individual-based models but also for continuous models formulated in terms of population sizes. Let us consider the example of an L-state system, in which there are L different linguistic groups, labeled as $\ell = 1, \ldots, L$, with respective population sizes $N_\ell(t)$ evolving with time according to the equations

$$\frac{dN_\ell}{dt} = \sum_{\ell'(\ell' \neq \ell)} [P(\ell|\ell')N_{\ell'} - P(\ell'|\ell)N_\ell] + rN_\ell \left(1 - \frac{\sum_{\ell'} N_{\ell'}}{K}\right). \qquad (7.18)$$

These equations describe transitions from group ℓ to group ℓ' taking place with rates $P(\ell|\ell')$, analogous to the two- and three-state models. Notice the form of the growth terms, where there is the same Malthus rate r for all groups and the equal expressions in the parentheses, depending on the ratio between the total population $N_{\text{tot}} = \sum_\ell N_\ell$ and the common carrying capacity K. This is just a specific example of growth, among many possible ones, in which the available resources are accessible to and equally shared among all the L populations. Summing Eqs. (7.18), one obtains that the total population N_{tot} evolves with time according to Verhulst dynamics,

$$\frac{dN_{\text{tot}}}{dt} = rN_{\text{tot}} \left(1 - \frac{N_{\text{tot}}}{K}\right),$$

which has the long time limit solution $N(t \to \infty) = K$. If this asymptotic solution $N_{\text{tot}} \approx K$ is replaced in Eq. (7.18), all the logistic terms vanish, formally turning Eqs. (7.18) into the corresponding population-conserving version. Thus, models with constant population can be used to study linguistic communities that (a) have reached an equilibrium total population size and (b) have equal access to resources. Such an approximation can be made, for example, in the study of the interaction dynamics of Castillan and Catalan in Spain or that of English and French in Canada. However, it could hardly hold in other situations, for example, in the following cases:

- competition and growth take place on the same time scale—in this case, population dynamics plays a crucial role;
- heterogeneous carrying capacities $\{K_\ell\}$, for example, a group ℓ may suffer a reduced access to resources characterized by smaller value of K_ℓ;
- resources are shared between the various groups—in this case, the carrying capacities can depend on the populations sizes (see the model of Kandler [2009] in Section 7.4.2.)

7.4.1 The model of Pinasco and Romanelli

A minimal model taking into account population dynamics was introduced by Pinasco and Romanelli (2006) (PR model). The PR model can be obtained from the AS model by adding logistic growth. In the neutral volatility case, $a = 1$, the model simplifies into a sort of basic Lotka–Volterra model with logistic dynamics for the two linguistic groups X and Y,

$$\dot{x} = c\,xy + r_X x\left(1 - \frac{x}{K_X}\right),$$

$$\dot{y} = -c\,xy + r_Y y\left(1 - \frac{y}{K_Y}\right), \qquad (7.19)$$

where $c > 0$ measures the switch rate from language Y to language X, assuming that Y is the minority language, r_X and r_Y are the Malthus rates, and K_X and K_Y are the carrying capacities, representing the asymptotic populations for isolated growth ($c = 0$). Pinasco and Romanelli (2006) showed that both languages X and Y can survive at the same time in a stable demography-induced equilibrium state as long as the minority language group grows fast enough, namely, faster than the rate at which its members transform into X individuals, which leads to the condition

$$cK_X < r_Y. \qquad (7.20)$$

If this condition is fullfilled, there is a stable mixed equilibrium state (x^*, y^*) with both population sizes $x^* > 0$ and $y^* > 0$, where the coordinates of the equilibrium point are given by

$$x^* = K_X \frac{1 + (cK_Y/r_X)}{1 + (cK_Y/r_X)(cK_X/r_Y)}, \qquad (7.21)$$

$$y^* = K_Y \frac{1 - (cK_X/r_Y)}{1 + (cK_Y/r_X)(cK_X/r_Y)}. \qquad (7.22)$$

7.4.2 Shared resources: The model of Kandler and Steele

As noticed by Kandler and Steele (2008), there is a problem in the stable solution of the model of Pinasco and Romanelli: the X equilibrium population size is larger than the corresponding carrying capacity, which is by definition, the maximum sustainable population: $x^* > K_X$, as can be seen from Eq. (7.21) considering that from the condition (7.20) one has that $(cK_X/r_Y) < 1$. The reason why $x^* > K_X$ is the continuous flux of individuals moving from Y to X and is the same reason why Eq. (7.22) gives an equilibrium Y population size $y^* < K_Y$.

Kandler and Steele (2008) introduced some novel models of population dynamics in which the population size remains limited below the expected threshold $x = K$. One of these models describes a common set of resources, measured as a total carrying capacity K (maximum number of individuals that can be sustained overall), which have to be shared among all the populations. This implies that in two-state models, the resources K will be shared between the X and Y groups that will have correspondingly different *time-dependent* carrying capacities: if the Y group has size $y(t)$ at time t, then for the X group, there remains an amount of resources sufficient to sustain a population not larger than $K_X(t) = K - y(t)$; similarly, an amount of resources $K_Y(t) = K - x(t)$ is available for the Y group at time t. In the version of the model corresponding to a fully connected network (or in which there is no spatial variable), the dynamical equations are as follows,

$$\dot{x} = c\,x\,y + r_X x \left(1 - \frac{x}{K-y}\right),$$

$$\dot{y} = -c\,x\,y + r_Y y \left(1 - \frac{y}{K-x}\right). \tag{7.23}$$

Here, the X↔Y transitions have been modeled as a Lotka–Volterra transition, proportional to the product of the population sizes xy like in the model of Pinasco and Romanelli, Eqs. (7.19). The main novelty of the model is in the logistic terms, which are different in the effective carrying capacities. The new growth terms in Eqs. (7.23) work as expected, modifying the dynamics of the model and limiting the population sizes below K, even if in this particular case of competition dynamics, they prevent the coexistence of X and Y and only one community can survive asymptotically. This model emphasizes the fact that the growth terms have a crucial role in the competition because they describe different underlying social and/or economical situations encountered in real-life cases.

7.5 Taking into Account Geography

It is known that geography is a relevant factor in the evolution and dispersal of species (Lomolino et al., 2006). Moreover in cultural diffusion, analogous factors exist, which can be relevant and have to be taken into account. Cultural diffusion can be influenced by many different factors related to historical events, the physical or political geography, the social structure of the society, the economic landscape and the underlying communication networks, reflected by different elements of the model, such as the initial conditions, the presence of boundaries, space- and time-dependent transition and growth rates, etc. As mentioned in Section 6.2.1, the evolutionary language dynamics in space produces regularities represented by relations between various quantities, such as language sizes, geographical extensions, and language diversity.

The evolution and spreading of languages is directly affected by various factors related to geography, from physical geography (terrain roughness, rivers, mountains, etc.) to the local economical geography and metereology. These factors are relevant for characterizing languages and for testing theoretical models of language dynamics, but also because the state of language at risk of extinction is very sensitive to such factors (Amano et al., 2014; Coelho et al., 2019; Xia et al., 2019). A detailed study of the influence of a wide class of environmental factors on the diversity of languages (Axelsen and Manrubia, 2014) has shown that rivers and the landscape roughness play particularly important roles.

In the following sections, we will discuss a few examples of extended models, constructed by generalizing their zero-dimensional counterparts (in this case, AS or MW model) by including space dimensions (Kandler and Steele, 2008; Kandler, 2009; Patriarca and Leppänen, 2004; Patriarca and Heinsalu, 2009).

To avoid confusion with the spatial coordinates, when dealing with space-dependent problems, the labels 'A' and 'B' for monlinguals and 'C' (or equivalently 'AB') for bilinguals will be used in place of X, Y, and Z, respectively. Correspondingly, the population sizes x, y, and z will be indicated now with n_A, n_B, and n_C and the corresponding population densities (number of individuals per unit area) with F_A, F_B, and F_C.

For the sake of simplicity, we skip the discussion of spatial models, such as the voter model or the AB model, in terms of physical analogies with domain growth in reaction–diffusion systems. For an interesting model of the dynamics of linguistic isoglosses, constructed on the analogy of surface tension-driven phenomena, see the model by Burridge (2017).

Let us start by considering the spatial version of the AS model. The straightforward spatial extension of the model (in the two-dimensional xy continuous space) can be done by promoting the ordinary differential equations of the AS model to partial differential equations, maintaining the form of the terms describing the A↔B transitions, and adding the diffusion terms modeling the dispersal of the populations. In practice, the differential equations (7.3) turn into the following reaction–diffusion equations,

$$\frac{\partial F_A(\mathbf{r},t)}{\partial t} = D_A \nabla^2 F_A(\mathbf{r},t) + R(\mathbf{r},t),$$

$$\frac{\partial F_B(\mathbf{r},t)}{\partial t} = D_B \nabla^2 F_B(\mathbf{r},t) - R(\mathbf{r},t), \qquad (7.24)$$

where $R = s_A F_A^a F_B - s_B F_B^a F_A$ is the A↔B reaction rate, $\mathbf{r} = (x,y)$ the position vector in the xy space, $\nabla^2 = \partial_x^2 + \partial_y^2$ the corresponding Laplacian, D_A and D_B the diffusion coefficients of the populations A and B, s_A and s_B the status of the languages, and a the volatility. Formally, these equations describe the diffusive motion and the A↔B reactions of a set of Brownian particles of two different types A and B (here interpreted as speakers or small groups of speakers of language A and B). The terms on the left-hand side represent the time variations of the local densities of speakers A and B, due to (1) the diffusion process across a two-dimensional space, and (2) the transformation process from A to B and vice versa, with rate R.

The functions F_i in Eqs. (7.24) represent densities, that is, the *local* average number of individuals per unit area in a small area at position $\mathbf{r}$ (and not the population sizes or the corresponding fractions). In the absence of population dynamics, the total population is conserved ($\int dx\, dy\, (F_A + F_B)$ = constant), as one can show by integrating in space the sum of Eqs. (7.24), and F_A and F_B can be normalized in order to represent *probability densities*, which are by definition globally normalized to one,

$$\int dx\, dy\, [F_A(\mathbf{r}, t) + F_B(\mathbf{r}, t)] = 1.$$

The fractions of A and B speakers at time t, represented in the following as $n_A(t)$ and $n_B(t)$, can be obtained from the corresponding probability densities by an integration over all the space, that is,

$$n_i(t) = \int dx\, dy\, F_i(\mathbf{r}, t). \quad (i = A, B),$$

and are correspondingly normalized, $n_A(t) + n_B(t) = 1$ at any time t.

It can be shown that the reaction–diffusion equations (7.24) have the same stable equilibrium states as the AS model. In the examples of models considered in the following sections, some features are added to the model, which reflect, for example, elements of a linguistic landscape and can have relevant consequences on the system evolution.

7.5.1 Inhomogeneous dispersal: Influence of linguistic barriers

A political or physical barrier may limit human dispersal, cultural exchanges, or both, from one side to the other side of the barrier. Here we consider the model introduced by Patriarca and Leppänen (2004), describing language competition in a two-dimensional square region divided into two regions α and β along the diagonal (see left panels in Figure 7.9) by a barrier. In this section, a barrier limits communication (and therefore the mutual cultural influence) between speakers located in the different regions, but does not affect the dispersal process, which takes place freely across the barrier. In practice, it affects the model in the part responsible for language change but not in the others (e.g., dispersal). We refer to this type of barrier as a 'linguistic barrier'. This is more a technical definition, related to the the terms of the equations which are modified, since there can be very different reasons causing the modulation of the rate of language shift. In this model, speakers are assumed to be initially divided into two groups: a fraction $n_A(0)$ of speakers of language A located in region α and the remaining fraction $n_B(0) = 1 - n_A(0)$ of speakers of language B located in region β, as shown in the left panels of Figure 7.9. The initial conditions for the A and B group are Gaussian probability densities localized in the respective region. The reaction–diffusion dynamics is built in analogy with that of the AS model, with two main differences:

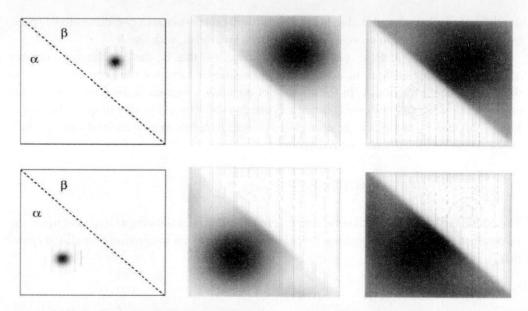

Figure 7.9 Plots of the probability densities F_A (lower panels) and F_B (upper panels) of A speakers and B speakers, respectively, at the initial time $t = 0$ (left panels), at an intermediate time $t = 200$ (central panels), and in the final stationary state $t = 10^4$ (right panels). The model parameters used are $a = 1.3$, $s_A = 0.52 = 1 - s_B$, and $D = 1$. The dotted line plotted in the left panels defines the border between regions α and region β. From Patriarca and Leppänen (2004) with permission from Elsevier.

- A homogeneous diffusion process takes place through the square domain $\{\alpha + \beta\}$ without any constraint, assuming reflecting boundary conditions on the sides of the square region.
- The A↔B transitions probabilities of an individual are only affected by the speakers in the same region (α or β).

The dynamics are defined as follows.

$$\frac{\partial F_A(\mathbf{r},t)}{\partial t} = D\nabla^2 F_A(\mathbf{r},t) + R(\mathbf{r},t),$$
$$\frac{\partial F_B(\mathbf{r},t)}{\partial t} = D\Delta^2 f_B(\mathbf{r},t) - R(\mathbf{r},t),$$
$$R(\mathbf{r},t) = P_{BA}(\mathbf{r},t)F_B(\mathbf{r},t) - P_{AB}(\mathbf{r},t)F_A(\mathbf{r},t). \tag{7.25}$$

Here, D is the common diffusion coefficient and P_{ij} are the rates of language shift. These equations are similar to those of the spatial extension of the AS model, Eqs. 7.24, but with the relevant difference that, as speakers diffuse across the square, they undergo $i \to j$ transitions with space rates and time-dependent rates $P_{ij}(\mathbf{r},t)$, depending only on the total number of speakers j in the same region to which the point $\mathbf{r}$ belongs, that is,

$$\begin{aligned} P_{AB}(\mathbf{r},t) &= s_B \left(n_B^{(\alpha)}\right)^a, \quad (\mathbf{r} \in \alpha) \\ &= s_B \left(n_B^{(\beta)}\right)^a, \quad (\mathbf{r} \in \beta) \end{aligned} \tag{7.26}$$

$$P_{BA}(\mathbf{r}, t) = s_A \left(n_A^{(\alpha)}\right)^a, \quad (\mathbf{r} \in \alpha)$$

$$= s_A \left(n_A^{(\beta)}\right)^a, \quad (\mathbf{r} \in \beta)$$

where the fractions $n_i^{(\alpha)}$ and $n_i^{(\beta)}$ of speakers i (i = A,B) in region α or β are

$$n_i^{(\alpha)} = \int_\alpha dx\, dy\, F_i(\mathbf{r}, t),$$

$$n_i^{(\beta)} = \int_\beta dx\, dy\, F_i(\mathbf{r}, t).$$

The model exhibits a stable equilibrium configuration, in which both languages A and B survive in the two regions α and β, respectively, for a wide interval of the language status.

The relevance of a spatial heterogeneity in the transition rates is clearly shown in a simple model introduced by Kandler and Steele (2008), which can be considered as the spatial extension of the model with shared resources discussed earlier (see Eqs. [7.23]),

$$\frac{\partial F_A}{\partial t} = D_A \nabla^2 f_A + c(\mathbf{r}) F_A F_B + r_A F_A \left(1 - \frac{F_A}{K - F_B}\right),$$

$$\frac{\partial F_B}{\partial t} = D_B \nabla^2 F_B - c(\mathbf{r}) F_A F_B + r_B F_B \left(1 - \frac{F_B}{K - F_A}\right), \quad (7.27)$$

where K now represents the maximum population density sustainable. The rate coefficient c is promoted here to a space-dependent function, $c \to c(\mathbf{r})$, in order to model two different spatial domains, a situation typically encountered whenever crossing a border marking the limits of two neighboring regions that differ in political, social, or economic aspects. The shape of the function $c(x)$ along the x-axis, schematized in the upper panel of Figure 7.10 (c is independent of y), is such that its sign favors language A in Region 1 and language B in Region 2. Notice that once the population dynamics are taken into account, the quantities F_A and F_B represent actual population densities and cannot be scaled to represent probability densities—the relevant scale of the population density variables in this case is the global carrying capacity K. According to results of the simulation of the model, both languages survive separated in the left and right regions with densities represented in the lower panel of Figure 7.10. Using asymmetric shapes of $c(x)$ or inhomogeneous diffusion coefficients $D_A \neq D_B$, one obtains asymmetrical solutions. The larger the values of the diffusivities, the larger is the overlap between the densities F_A and F_B.

We also mention here a more general model introduced by Kandler (2009), that takes into account bilinguals, besides dispersal (through the diffusion terms) and social structure (through the growth terms representing shared resources). Indicating by C≡{AB} the bilingual status, the coupled equations for the two monolingual population densities F_A and F_B and the bilingual population density F_C are as follows.

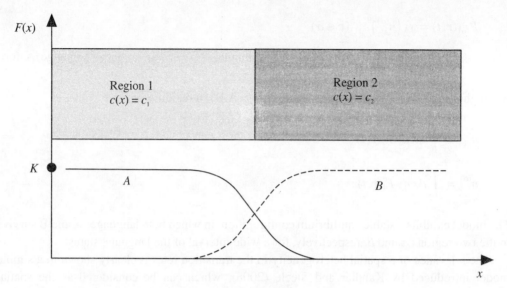

Figure 7.10 Equilibrium solution for the population densities F_A and F_B in the two-dimensional model described by Eqs. (7.27), in which the transition rate $c = c(x)$ is x dependent and diffusion is homogeneous, $D_A = D_{AB}$. Adapted from Kandler and Steele (2008).

$$\frac{\partial F_A}{\partial t} = D_A \nabla^2 F_A + r_A F_A \left(1 - \frac{F_A}{K - F_B - F_C}\right) + k_{CA} F_A F_C - k_{AC} F_B F_A,$$

$$\frac{\partial F_B}{\partial t} = D_B \nabla^2 F_B + r_B F_B \left(1 - \frac{F_B}{K - F_A - F_C}\right) + k_{CB} F_A F_C - k_{BC} F_A F_B,$$

$$\frac{\partial F_C}{\partial t} = D_C \nabla^2 F_C + r_C F_C \left(1 - \frac{F_C}{K - F_A - F_B}\right) - (k_{CB} F_A F_C + k_{CA} F_A F_C)$$
$$+ (k_{BC} + k_{AC}) F_A F_B. \tag{7.28}$$

This model was applied to the case of Celtic languages for designing possible support strategies (Kandler et al., 2010). The model was also tested assuming heterogeneous diffusion (Kandler, 2009), that is, when the populations have different diffusion coefficients. We notice that heterogeneous diffusion is a necessary condition for the appearance of Turing patterns (Murray, 2003) but so far Turing patterns have not been reported in any study of language dynamics.

This general model suggests in turn a series of interesting variants, depending on the underlying linguistic situation. For example, the transition rates can be made space- and time-dependent as for the model described by Eqs. (7.27) by promoting the rate coefficients to functions of position and time. This change is expected to lead to novel final scenarios where, for example, monolinguals and bilinguals can coexist in separate linguistic islands. Furthermore, the current form taken from the MW model could be replaced by one of the types of interactions of the models of bilinguals discussed earlier.

Finally, the growth terms describing shared resources lend themselves easily to describe different social and economic structures. For example, the form of the growth terms in Eqs. (7.28) is symmetrical in the three populations, implicitly assuming that the three groups A,

B, and C access resources in different ways in competition with each other. In a situation where bilingualism is driven by the economic motivation of economically joining group A, it is possible that bilinguals C and the monolinguals of the higher status language A form a homogeneous group enjoying similar resources, while the monolinguals of the minority language B are described by a diversified growth term.

A further study in Kandler (2009) considers *nonlocal dispersion*, by replacing the Laplacian operator in Eqs. (7.28) with a non-local integral operator. Non-local diffusion and anomalous diffusion processes are seldom considered in language dynamics (see Sofuoglu and Ozalp [2017] and Owolabi and Gómez-Aguilar [2018] for for interesting exceptions). The same non-local extension that can be introduced in the diffusion process can also act in the linguistic interactions, that is, in the competition terms describing the transitions, for which there is a vast literature of ecological competition models characterized by non-local competition with finite competition ranges.

7.5.2 Inhomogeneous dispersal: Influence of physical geographical barriers

In a spatial two-state system, the introduction of space-related constraints may have relevant consequences at the macroscopic scale, such as changing the final equilibrium state reached by the system. These effects can be caused, for example, by changing the boundary conditions (e.g., introducing a reflecting boundary) or when some small changes are induced in the initial conditions of the populations densities (see Patriarca and Heinsalu [2009] for some examples). In this section, we consider another example of space-induced effect, in a way complementary to the linguistic barrier discussed in Section 7.5.1. We will consider the effect produced by the presence of a 'physical geographical barrier', that is, a region of space which modulates the drift and/or diffusion process (caused, e.g., by the underlying geographical landscape) that leaves the linguistic dynamics responsible for the A↔B switching unaffected (Patriarca and Heinsalu, 2009).

The example considered is a close ecological analog of a competition and selection process between two different species A and B, initially located in two different islands α and γ, trying to colonize at the same time a previously desert island β located between α and γ. Diffusion is basically free inside each island, but the probability of diffusing from an island to another one depends on — and decreases with — the distance between the two islands. The initial conditions and the geography of the landscape are depicted in the upper panel of Figure 7.11, in which the A group, with a higher language status $s_A = 0.6$, is initially located entirely in the island α, while the B group, with $s_B = 0.4$, is only present in island γ. The goal of the numerical experiment is to show that, in spite of the higher status of language A (competition advantage of species A from an ecological perspective), the smaller distance from island β to γ than to α allows a successful colonization of β by the B group.

The inhomogeneous diffusion process is modeled on the analogy of drifted Brownian motion of a Brownian particle in an external potential $U(x,y)$ (van Kampen, 1992). The potential U is chosen with a suitable shape (shown in the lower panel of Figure 7.11) for mimicking the diffusion process across the island landscape. Each island is modeled as a

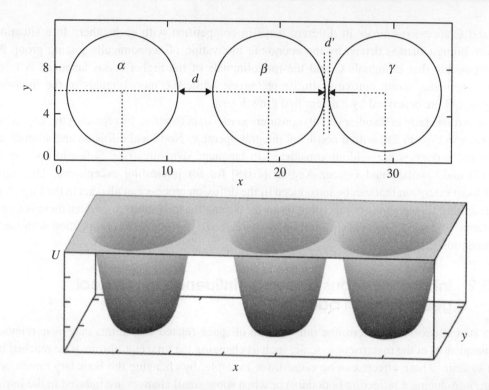

Figure 7.11 Top panel: Schematic map of the three islands α (left), β (center), and γ (right). Lower panel: Potential shape $U(x, y)$ used in the Brownian motion model of Eqs. (7.29). The distance d' between island β and island γ is smaller than the distance d between islands α and island β. The A and the B groups are initially present only on island α and β, respectively. See text for further details. Reprinted from Patriarca and Heinsalu (2009) with permission from Elsevier.

potential well and an individual migrating from an island to another is equivalent to the Brownian particle crossing the corresponding potential barrier (see Patriarca and Heinsalu [2009] for details).

The dynamics is defined by the following reaction–diffusion equations,

$$\frac{\partial F_A}{\partial t} = R(F_A, F_B) - \nabla \cdot (\mathbf{f} F_A) + D \nabla^2 F_A,$$

$$\frac{\partial F_B}{\partial t} = -R(F_A, F_B) - \nabla \cdot (\mathbf{f} F_B) + D \nabla^2 F_B,$$

$$R(F_A, F_B) = s_A F_A^a F_B - s_B F_B^a F_A. \tag{7.29}$$

The form of the reaction rate $R(F_A, F_B)$ is similar to that of the AS model; the terms proportional to D are diffusion terms, while the remaining terms describe advection under the action of the force $\mathbf{f}(x, y) = (f_x(x, y), f_y(x, y)) = -\nabla U(x, y)$. In this case (homogeneous advection and dispersal), the total probability density $F = F_A + F_B$ follows a standard diffusion–advection process, which is obtained by summing Eqs. (7.29),

$$\frac{\partial F}{\partial t} = -\nabla \cdot (\mathbf{f} F) + D \nabla^2 F. \tag{7.30}$$

Numerical simulations show that if the distance between α and β is equal to that between β and γ, language A (the language with higher status) colonizes island β, as expected. If the position of island β is gradually moved closer to island γ, barrier crossings for an A individual from island α to β becomes more and more difficult, with the consequence that the density of A speakers in island β will be smaller and smaller, while that of B individuals will increase. There is a critical β–γ distance at which the density of A speakers is not sufficient anymore to ensure final colonization following a local AS dynamics and the B group eventually prevails.

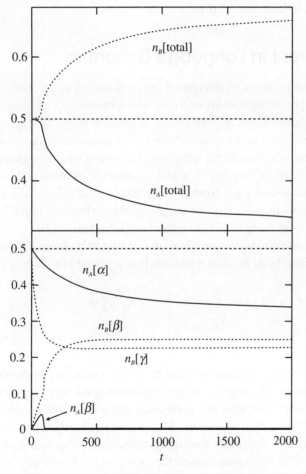

Figure 7.12 Population fractions n_A and n_B versus time obtained from the numerical solutions of Eqs. (7.29). Top panel: Total fractions of speakers A and B in the system (notice that both A and B survive asymptotically). Bottom panel: Fraction of speakers A and B in each island α, β, and γ (island indicated in parentheses), the dashed line $n = 0.5$ marks the initial population sizes. (The fractions $n_A[\gamma]$ and $n_B[\alpha]$ are not visible because they are always negligible throughout the simulation.) Reprinted from Patriarca and Heinsalu (2009) with permission from Elsevier.

The larger distance makes island β a *refugium* for the B group, since there is an advantage for the B group in accessing island β.

The time dependence of the various population fractions is shown in Figure 7.12 for a geometrical configuration of the system leading to the final colonization of island β by the B group. Notice that the A group was present for some time in island β ($n_A[\beta] > 0$) but then the population is absorbed by the B group. The fractions $n_A[\gamma]$ and $n_B[\alpha]$ are always negligible throughout the simulation.

This phenomenon happens not only in two-state competition models like the spatial AS model, but also for the spatial extensions of three-state models, such as the MW model. In the latter case, one can find, besides equilibrium configurations where the single languages survive, also the other interesting possibility that the bilingual group survives alone in a suitable refugium, as shown in Giordano (2015).

7.6 Wave Front in Language Dynamics

In this section, we will discuss an interesting question arising in the study of language shift in space, namely, the propagation of wave fronts. As a language X gradually displaces another language Y, the linguistic landscape can evolve in different ways, depending on the underlying dynamics. In particular, it can happen—both in the presence or the absence of population dynamics—that there is a wave front, separating the region where language X is spoken from the corresponding region of language Y, which is observed to move with a certain velocity v^*. The conditions under which wave front propagation is observed and the properties of such a propagation is both a mathematically challenging problem (Fort and Pujol, 2008) and a question of general interest in the study of culture spreading (Isern Sardó, 2011).

Within language dynamics, this problem was addressed by Isern and Fort (2014) through a novel model, described by dynamical equations that we write in a slightly more general form,

$$\frac{\partial F_A}{\partial t} = D_A \nabla^2 F_A + r F_A \left(1 - \frac{F_A + F_B}{K}\right) + \gamma_A F_A^\alpha F_B^\beta \Phi_{\alpha\beta} - \gamma_B F_B^{\beta'} F_A^{\alpha'} \Phi_{\alpha'\beta'},$$
$$\frac{\partial F_B}{\partial t} = D_B \nabla^2 F_B + r F_B \left(1 - \frac{F_A + F_B}{K}\right) - \gamma_A F_A^\alpha F_B^\beta \Phi_{\alpha\beta} + \gamma_B F_B^{\beta'} F_A^{\alpha'} \Phi_{\alpha'\beta'}, \quad (7.31)$$

where D_A and D_B are the diffusion coefficients and a common Malthus rate r and carrying capacity K are assumed. The reaction terms, representing the language shifts B↔A, resemble those of the AS model, but with the population densities raised to the powers $\alpha, \alpha', \beta, \beta'$, playing the roles of four independent volatility parameters. Furthermore, Isern and Fort (2014) assume that (a) transition rates depend on the probabilities (rather than on the actual densities), hence the presence of the re-normalization factor $\Phi_{\alpha\beta} \equiv 1/(F_A + F_B)^{\alpha+\beta-1}$, and (b) the A→B transitions from the higher status language A to the lower status language B are negligible, setting $\gamma_B = 0$. The AS dynamics is recovered for $\alpha = \beta' = a, \beta = \alpha' = 1$, and $\Phi_{\alpha\beta} = 1$.

Notice that in the example of inhomogeneous diffusion of Eqs. (7.29), the rates depend on the absolute values of the population densities. Which possibility is more realistic depends on the system considered: using probabilities instead of densities means that we need to assume that the transition probabilities depend on the average properties of the *whole* system—which implies a well-connected system in which information can flow fast; on the contrary, using densities means to assume that changes can only depend on the local properties of the system, corresponding to a system in which the information flow is slow and limited between first neighbors. The real-life scenario is probably halfway between these two choices, since it is known that the structure of social networks, through which information and culture flows, is neither a perfect lattice nor a fully connected network, but has small-world properties. A study based on space-embedded complex networks may be a valuable approach to culture and language spreading.

In order to determine the values of the coefficients γ_A, α, and β, Isern and Fort use a non-spatial version of the model described earlier to make a best fit of the data in the database of Quechua, Welsh, and Gaelic, already considered in the AS model. The results of this analysis are also interesting because the authors obtain values of α and β which differ for each database analyzed, thus questioning the suggestions of Abrams and Strogatz that there may exist an average, universal value of such exponents (see also Sutantawibul et al. [2018] for a critical reanalysis). With the help of some estimates for the remaining parameters, the front velocities v^* from the various databases are found to have values in the interval $v^* \in (0.3, 0.9)$ km/year.

7.6.1 The ecology of languages: The Baggs and Freedman model

We conclude this selection of competition models discussing—to the best of our knowledge—the first language competition model, introduced by Baggs and Freedman (1990); Baggs et al. (1990); Baggs and Freedman (1993). The works of Baggs and Freedman were brought to the attention of the complex systems community by Kandler and Steele (2008) and were reviewed by Kandler (2009). The Baggs and Freedman (BF) model is a remarkable model of language dynamics. Not only does it predate most of the later works of language competition in many relevant points, but it also contains some noteworthy features that have not been fully studied or considered yet in other models of language dynamics. The model introduced by Baggs and Freedman (1990); Baggs et al. (1990) is a two-state model describing the interaction between the minority language group and the bilingual group, which was revisited and analyzed more recently by Wyburn and Hayward (2008) and applied to various real situations. The model by Baggs and Freedman (1993) is a three-state model, taking into account the full dynamics of both monolingual groups A and B and of the bilingual group C; it was further studied and extended by Wyburn and Hayward (2009, 2010). The richness of the model makes it an attractive framework for analyzing real-life situations, since it can generate a large variety of equilibrium scenarios—this is the reason why we are discussing it here rather than at the beginning of the chapter.

We limit ourselves to consider the general model introduced by Baggs and Freedman (1993), which is a three-state model describing two monolingual groups X and Y and the bilingual group Z (since the BF model is a zero-dimensional model, in this section, we go back to the notation X, Y, and Z to indicate the two monolinguals groups and the bilinguals, respectively, without creating ambiguities). For the sake of clarity, we do not consider the most general form but discuss one of the explicit examples considered by Baggs and Freedman (1993), in which the dynamical equations of the population sizes $x(t)$, $y(t)$, and $z(t)$ can be written in the following way,

$$\frac{dx}{dt} = r_X x - \gamma_X x^2 + p_X z - k_X \frac{xz}{1+x+z}\left(1 + \frac{k'_X y}{1+y}\right),$$

$$\frac{dy}{dt} = r_Y y - \gamma_Y y^2 + p_Y z - k_Y \frac{yz}{1+y+z}\left(1 + \frac{k'_Y x}{1+x}\right),$$

$$\frac{dz}{dt} = (r_Z - p_X - p_Y)z - \gamma_Z z^2$$
$$+ k_X \frac{xz}{1+x+z}\left(1 + \frac{k'_X y}{1+y}\right) + k_Y \frac{yz}{1+y+z}\left(1 + \frac{k'_Y x}{1+x}\right). \tag{7.32}$$

The first terms on the right-hand side of the equations can be recognized as logistic growth terms (r_i and γ_i are constants, $i = X, Y, Z$, they also contain the contribution of emigration), which represent the contributions due to *vertical* transmission (from parent to offspring), while the other terms (proportional to k_X and k_Y) are those representing horizontal transmission. Notice that the Malthus rate r_Z for the bilingual population z is re-normalized by the negative quantities $(-p_X)$ and $(-p_Y)$, representing the fractions of children born in a bilingual family but growing up monolingual in language X (p_X) or in language Y (p_Y); correspondingly, there are equal positive contributions in the equations of x and y—the condition $r_Z - p_X - p_Y > 0$ has to hold (see Baggs and Freedman [1993] for further details).

In order to interpret the horizontal transmission terms, we set $k'_X = k'_Y = 0$ for the moment. Then, the expressions inside the parentheses are equal to 1 and the terms can be recognized as representing monolingual→bilingual transitions with transition rates per individual given by

$$P(Z|X) = p_X z, \qquad P(X|Z) = \frac{k_X z}{1+x+z},$$
$$P(Z|Y) = p_Y z, \qquad P(Y|Z) = \frac{k_Y z}{1+y+z}. \tag{7.33}$$

The transition rates $P(Z|X)$ and $P(Z|Y)$ in the BF model have the same form as in the MW model. Notice the dependence of the $P(X|Z)$ and $P(Y|Z)$ transition rates on the total number of bilinguals z rather than on the Y population size, as in the MW model. This is equivalent to the assumption that monolinguals know/hear their language only via the bilinguals (but do not communicate in any way with monolinguals of the other language). The additional factors $1/(1 + x + z)$ in the equation for $x(t)$ and $1/(1 + y + z)$ in the equation for $y(t)$ are 'Holling factors', used in ecological competition models for re-normalizing the transition rates through the population sizes (Freedman, 1980; Hastings, 1996). This correction, usually neglected in language dynamics, is important for avoiding arbitrarily large horizontal transmission rates. For example, for a fixed x, the X→Z rate in the absence of the Holling factor would be proportional to z and therefore, would become unrealistically large if the bilingual population size z would grow enough.

The terms with the coefficients k'_X and k'_Y introduce a novel feature, that is, a three-individual (rather than a pair-wise) interaction. This is signaled by the three-population product $\dot{x}, \dot{y} \propto x \times y \times z$, which can be interpreted as the rate of encounters between a X speaker, a Y speaker, and a bilingual speaker Z (the expressions $1/(1+y)$ and $1/(1+x)$ are the Holling factors regularizing the transition rates). Effects of this type, which can be referred to as 'third-order effects', are usually not taken into account in most basic language dynamics models. A three-agent meeting will contribute—at least in a fraction of the events—to the enhancement of the X→Z and Y→Z transitions. In fact, a dialogue between a Y speaker and a bilingual Z can only take place in language Y. A speaker X will meet a pair of individuals Y and Z at the same time—and assist in a dialogue in the language Y—with a probability proportional to the expression xyz. Similar considerations can be done for an individual Y meeting a bilingual Z and a monolingual X talking to each other.

Noteworthy exceptions, in which third-order effects are taken into account, are the models of language use mentioned in Section 7.3.6, which focus on the conversation processes between individuals rather than on the transitions. In fact, in the model by Carro et al. (2016), triangular structures of the social network (associated with three-individual-based interactions) play a crucial role. The model used by Karjus and Ehala (2018) for analyzing the social complexity of a database of bilingualism in Estonia is also based on the use of triadic closure. Moreover, in the model of Iriberri and Uriarte (2012), which is focused on language use, the effective mean field processes associated with the game theoretical dynamics can be shown to be described by nonlinear differential equations containing transition rates depending on the third power of the population sizes.

The rich nonlinear dynamics of the BF model leads to six equilibrium points in general, representing the different possible scenarios in terms of survival and extinction, coexistence, and persistence. Besides introducing a model that can still be considered novel and challenging in many features, Baggs and Freedman also raised some questions that would be worth studying in order to provide a description of real-life situations (Baggs and Freedman, 1993):

- The carrying capacities of each population may not be independent of each other. An example of sharing of the available resources is the model of Kandler and Steele (2008).
- People continuously move in or out of a given community. This can suggest many research directions to extend the model, in particular, generalized models including dispersal, a feature considered by models with spatial extensions.
- The age structure of the population has an important role, a question seldom considered in language dynamics models (see Schwämmle [2005] for one of the few studies on language dynamics with aging individuals).
- There is a finite time scale associated with the learning process of children, which can be taken into account by some models with time delays.

All in all, language dynamics is beginning to undertake the study along the main lines started or foreseen in the seminal work of Baggs and Freedman.

7.7 Examples of Language Use

This section will focus on the phenomenological aspect of language competition as part of language dynamics in space and time, from a twofold standpoint—qualitative and quantitative—based on raw data (official statistics about domains of use of languages in contact). As such, it provides a diversified and concrete overview of a situation similar to those we have modeled or simulated in the previous section.

However, we do not make a simulation-based study for reproducing the data, but limit ourselves to some general considerations: this study is intended as an intermediate step toward the development of a model. For an example of a language competition model-based analysis of a linguistic database, see the detailed analysis of Karjus and Ehala (2018), where the results of the survey 'Linguistic attitudes in Estonia 2015' (Karjus and Ehala, 2015) are considered. There are other databases that can serve equally well for this aim; all of them represent open challenges to the language dynamics of language evolution, competition, and use.

We shall handle domain-anchored data in two typical bilingual or multilingual settings, in a situation of typological discontinuity (Basque) and structural continuity (Catalan), as compared to the dominant official language (Spanish). In both cases, the previous situation, before additive language planning started in the early eighties of the past century, was akin to Fishmanian and Fergusonian diglossia, respectively.

The linguistic database used is the EAS database of the Basque dialects, *Euskararen Atlas Sozio-geolinguistikoa* ('Socio and Geo-linguistic Atlas of the Basque Language').[3]

In the following, X refers to the monolingual speakers of the minority language, Y refers to the monolingual speakers of Castilian, the official language in Spain, and Z to bilingual speakers. Index chains read as follows: the first X,Y, or Z index points at the context or the dominant or prevailing settings; whereas, the second index hints at the emerging or enhanced trend; for example, XZ can represent bilingualism rooted in Basque current or vernacular use; YZ a bilingualism draining Spanish monolingual or Erdara speakers to its sphere; etc. Furthermore, the symbol '>' means a change in a trend of linguistic use, for example, XZ > YZ means a passage from XZ to YZ settings.

7.7.1 Language use in the Basque country

Hints from domains and scope of use

Léonard (2007) provides a study of the sociolinguistic patterns in the years 1980–1990 in the Basque country, based on the following assumptions:

  (i) Linguistic policies on behalf of additive bilingualism foster pluralism and integration in society, and can be used as an efficient tool for peaceful and viable nation-building.

---

[3] Euskararen Atlas Sozio-geolinguistikoa ('Socio-Geolinguistic Atlas of the Basque Language' (EUDIA).
Home Page: https://sites.google.com/site/edakeudia/Home/aurkibidea/eas.
Collaborators: https://sites.google.com/site/eudiaehu/home/partaideak.
Institutions:
– Universidad del Paìs Vasco-Euskal Herriko Unibertsitatea (UPV-EHU);
– Université de Pau et des Pays de l'Adour (UPPA);
– Ministerio de Ciencia e Innovaciòn (MICINN).

(ii) Statistical clues of emerging additive bilingualism on the one hand (in the southern part of the Basque country, in Spain) and of subtractive bilingualism (in the French area, or northern part of the Basque country) on the other hand should be analyzed according to a consistent model of integration, such as that provided by Karklins (2000); Lauristin and Heidmets (2002), that is, *integration theory*, as emerged in the 1990s in the Baltic states.

(iii) The data provide useful information on the intricate sociocultural patterns from the standpoint of social innovation (additive bilingualism) and resilience or consensus out of a segregated situation (within subtractive bilingualism).

The former model of society hinted at patterns of separation and pluralism, whereas the latter enhanced patterns of assimilation and segregation, though heading toward another type of change. Using the three-state model notation, we can say that the tighter sociolinguistic setting, favoring the transitions Z→Y, is replaced by a more symmetrical X→Z / Y→Z dynamical situation.

Three main factors of overall integration of social sectors, adhering to X, Y, or Z, were distinguished:

- A structural factor—laws and citizenship.
- A functional factor—work sphere and sociability.
- An attitudinal factor—psychosocial and political; attitudes and opinions.

Our standpoint is different (though still relevant with integration theory) as we make an attempt to provide cues to modelers in order to read and interpret variables and parameters to project these categories and weights into models and simulations.

Table. 7.1 provides an overview of the use of the Basque language, spanning the 20-year long period between 1991 and 2011. From the table one can extract data on the relative frequencies of usage of the Basque language. The use of the minority language at stake (X) at home in 1991 matches $T^1(i)$ for 16.6% of the overall population, whereas the use of X with children according to the same census matches T^1 (j) for 17% versus 15% with friends

Table 7.1 Percentages of users of Basque (with respect to Spanish) among the population of the Basque Country with age ≥ 16 years for different scopes of use (= i, j, k, l, m, n) and years (1991, 2001, 2011). Source: Basque Government. Department of Education, Political Linguistics and Culture. Sociolinguistic Survey.

Scope: Use of Basque with	Scope label	T^1(1991) XZ > ZY	T^2(2001) ZY > YZ	T^3(2011) YZ > …	Micro-models (generic trends)
Family at home	i	16.6	17.4	17.2	XZ
Children	j	17.0	21.6	22.8	ZY > YZ
Friends	k	15.0	18.4	21.4	XZ > YZ
Coworkers	l	15.0	18.4	24.4	XZ > YZ
Local government workers	m	14.0	20.4	24.7	(Z)Y > YZ
Health service workers	n	8.0	14.2	19.8	(Z)Y > YZ

$T^1(k)$, etc. T^1 is symbolically described as an XZ > ZY initial state with a direct reference to the models of bilingualism considered above, meaning that this phase strongly inherits from the lower status period (the Fischmanian diglossic period) of subtractive bilingualism, when Spanish was the sole official language in the Basque region. Most Basque speakers had already become bilinguals, a situation described as XZ, referring to the correspondingly expected high rate of linguistic transformation. The ZY model suggests that this pattern of bilingualism eventually tends to foster Erdara monolingualism. The diglossic trend heads toward a language shift for this sector of the population; it is represented by ZY, meaning a dominant assimilation process. The small cohort of population concerned (less than 20%) flows from the previous coercive policy of assimilation, which shows up clearly from the $T^1(i)$ score of 16% of the overall Basque population declaring they used to speak X at home. Nevertheless, the values of T^1 refers to a situation occurring ten years after the beginning of consistent policy on behalf of additive bilingualism, which explains the score $T^1(j) = 17\%$; it has become an asset to be bilingual, that is, the XZ trend is on the rise and it becomes legitimate to encourage children to become bilinguals, fostering a ZY > YZ transformations. Nevertheless, X still follows diglossic patterns of vernacular use, as all values rise out of the most vernacular situation $T^1(i)$ especially, which shows either stagnant or decreasing trend in the use of Basque. The YZ model points at an incipient drift toward bilingualism (Z).

As showed in Table 7.2, while $T^1(j)$ tended to favor a ZY > YZ trend in terms of the micro-model of bilingualism, increasing conditions of additivity by 1%, now $T^1(k-l)$ point rather at an XZ > ZY trend, representing a model of assimilation, whereas $T^1(m-n)$ straightforwardly hints at an XZ > ZY trend for $T^1(m)$ and an XZ > Y one for $T^1(n)$, that is, one of drastic assimilation or acculturation of the most formal social setting. This model is induced by the thread of the speaker networks: not all civil servants nor health service staff can speak Basque, especially in the incipient phase of the language planning policy, which considerably lowers the score (14 and 8, respectively).

Table 7.2 Micro-models in the T^1 situation (1991).

Scope: Use of Basque with	Scope label	T^1 (1991) XZ > ZY	Micro-model (trends)
Family at home	i	16.6	XZ
Children	j	17.0	ZY > YZ
Friends	k	15.0	XZ > ZY
Coworkers	l	15.0	XZ > ZY
Local government workers	m	14.0	XZ > ZY
Health service workers	n	8.0	XZ > Y

Now, the T^3 situation in Table 7.3 shows a drastic reversal of these trends after three decades of policy of additive bilingualism. All domains of use show increasing values of bilingualism, pointing at a trend toward XZ and YZ, according to our qualitative interpretation of these

Table 7.3 Micro-models in T^3 situation (2011).

Scope: Use of Basque with	Scope label	T^3 (2011) YZ > ...	Micro-model (trends)
Family at home	i	17.2	XZ
Children	j	22.8	XZ/YZ > Z
Friends	k	21.4	XZ/Y > YZ
Coworkers	l	24.4	XZ/Y > YZ
Local government workers	m	24.7	Z(Y) > YZ
Health service workers	n	19.8	(Z)Y > YZ

results: while home ($T^1(i)$) remains the core situation for X and does not increase much, as it highly depends on Malthusian demographic increase, all other cohorts display drastic bursts uphill due to trends such as XZ/YZ > Z for $T^1(j)$, with 22.8%, as we have now to account for at least two cohorts—children raised in Basque since the last decades of the twentieth century who talk in Basque to their own children XZ/YZ > (Z), and new speakers of Basque talking Basque to their children YZ > Z. Both cohorts raise the total amount of bilinguals ($\rightarrow$ Z). As to $T^1(k-l)$, we observe two cohorts: genuine Basque speakers talking Basque to their friends and coworkers XZ/Y > YZ, but also monolinguals attracted by the bilingual model YZ, involved in a Y > YZ process of Basquization. Paradoxically, as compared to a diglossic situation, the impact of the sub-cohort of monolinguals can be seen especially in the group of coworkers for $T^1(l-m)$, with scores of no less than 24.4% and 24.7%. The confluence of 'old speakers' (euskaldun zaharrak) with 'new speakers' (euskaldun berriak) sub-cohort inflates the score here, whereas the value drops for the cohort of speakers using Basque in health service workers as this sector highly depends on a specific and rather mobile manpower, among which speakers of the Y are more likely to be involved in interactions.

Figure 7.13 shows the impact of language management or planning in a communal aggregate where the initial trend resorted to assimilation after a transitional phase of bilingualism (XZ > ZY). Moreover, these data show how the situation evolved within the span of three decades (1981–2011). The impact of language planning is twofold: on the one hand, a wider social practice of X is encouraged and actively fostered by bilinguals (the Z group adheres more actively to X and recruits more and more members); on the other hand, the use of X strongly depends on the XZ population, who plays a leading role as language mentors and models to the YZ cohorts. As a result, the T(i) values remain rather stable, whereas drastic increases occur in scopes of use where the YZ and Y type of populations are involved. In other words, although the former cohort of speakers (euskaldun zaharrak) is a closed set, whereas the latter (euskaldun berriak) is an open set, the former has a huge impact on the latter.

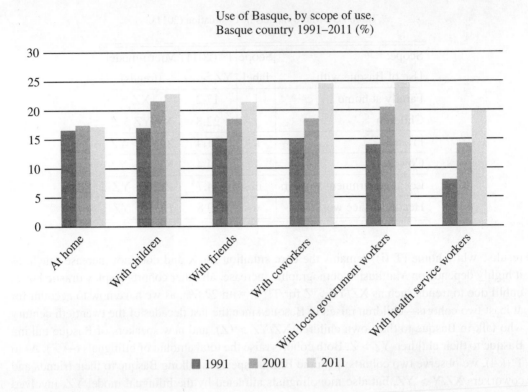

Figure 7.13 Use of Basque as much as or more than Spanish among the population of the Basque Country (with age ≥ 16 years) by the scope of use according to province, 2011.

Regional variation in emerging models of language competition

Table 7.4 displays interesting cues on the incidence of an additive linguistic policy, according to regions of the southern Basque country: the province of Araba/Alava is where the Basque government is located, providing a conspicuous array of working positions in public administration. As most of the population is monolingual in Spanish, although with a blend of incipient bilingualism, it can be described as a typical YZ situation. Gipuzkoa typically resorts to a XZ situation, as the region where Basque has a strong hold, with a dense network of small towns, villages, and hamlets where Basque is intensively spoken alongside with Erdara (i.e., Spanish). This zone strongly resisted the diglossic policy during Franquism; it can be considered as a citadel for X.

Bizkaia stands in the middle as some sub-regions have been nearly totally acculturated before the democratic transition, for example, the town of Bilbao, which can be considered as the main Basque town, whereas the coast and rural areas or small towns still used Basque in the nineteen eighties, when the additive policy was implemented. One could classify this region as having undergone a ZY process. In short, the main trends in the three regions read roughly as follows in terms of micro-models:

- Alaba: YZ,
- Bizkaia: ZY,
- Gipuzkoa: XZ.

Table 7.4 shows striking figures for Alaba, where YZ patterns dominate the distribution: the variable $Z_Y(i)$ (at home) suggests that most speakers are new learners (or 'new speakers', i.e., euskaldun berriak), or that households tend to be mixed. Instead, Z_Y(k-m) inverses the classical diglossic pattern: X is more spoken everywhere else than at home, and especially in the work sphere, where it reaches a peak (10.1%). $Z_Y(j)$ stands somehow apart, as a cue to how much X is conceived as an asset for children. Z_Y(k,m) hints that speakers tend to train as much as they can in Basque, though they have to cope with an overall low level of bilingualism—a problem of (critical) mass and density of Z network, notwithstanding the nearly total lack of X speakers in this region, artificially promoted in the capital of the Basque Country in order to preserve regional sociopolitical and cultural balance.

Table 7.4 Use of Basque among the population of the Basque Country with age ≥ 16 years by scope of use, according to province, 2011.

Scope (2011): Use of Basque with	Scope label	T^3 ARABA Initial state: YZ	T^3 BIZKAIA Initial state: ZY	T^3 GIPUZKOA Initial state: XZ
Family at home	i	3.8	11.6	31.9
Children	j	7.1	16.4	39.9
Friends	k	7.6	14.1	39.4
Coworkers	l	10.1	18.0	40.7
Local government workers	m	7.9	16.3	45.9
Health service workers	n	4.3	11.8	39.9

Table 7.5 points at the intricacy of these fluctuations, from the K(i) situation, in which monolingual speakers of the Y type hardly behave as bilinguals (a paradoxical situation summed up by a Y(Z) micro-model), while they tend to foster a more proper YZ model for K(j,k,m) scopes. It is an attempt at reaching a (Y)Z state in the K(l) pattern, which looks typically like a community of practice through career-driven incentives, as the Z model happens to be most rewarding for public servants, in a region where government jobs happen to be the main resource. At the other extreme of the polarity, in the K(n) scope, they face the same lack of speakers as at home so that the relevant micro-model could therefore be described as Y(Z).

A much more consistent overall situation prevails in the province of Gipuzkoa (Table 7.5), corresponding to a stable XZ model, which can hardly be mitigated or specified: it confirms the pivotal role of Basque at home (31.9%), and increases in all more formal scopes. The main lesson, though, is that it seems a critical mass has been reached here, which makes it possible for Z to gain a rather good stronghold in the more official settings, such as K(m-n).

Then, the main situation for the intermediate case in this polarity between the Alava career-driven YZ trend and the Gipuzkoa XZ model appears in Table 7.6, with the province

Table 7.5 Micro-models by scope of language use, Alava, 2011.

Scope (2011): Use of Basque with	Scope label	T^3 ARABA Initial state: YZ	Micro-models (trends)
Family at home	i	3.8	Y(Z)
Children	j	7.1	YZ
Friends	k	7.6	YZ
Coworkers	l	10.1	(Y)Z
Local government workers	m	7.9	YZ
Health service workers	n	4.3	Y(Z)

of Bizkaia. The situation of the K(i) variable shows that a strong trend to acculturation and attrition has been taking place for decades, though speakers are working hard at reversing this drift, ZY > YZ. Most other values suggest the same processes are evolving as in Alava, although the situation is much more variegated. More fine grained results would be needed here, in order to distinguish between the Bilbao situation and Basque resilience and normalization in rural settings and small towns.

Table 7.6 Micro-models by scope of language use, Bizkaia, 2011.

Scope (2011): Use of Basque with	Scope label	T^3 BIZKAIA ZY > Y/XZ	
Family at home	i	11.6	ZY > YZ
Children	j	16.4	YZ
Friends	k	14.1	YZ
Coworkers	l	18.0	(Y)Z
Local government workers	m	16.3	YZ
Health service workers	n	11.8	(Y)Z

Regional and gender variation

The next set of data allows us to compare a situation of additive bilingualism in the *Basque Autonomous Community* (CAB), or the Euskadi Community in Spain, or Hegoalde, with a substractive one in the *French Basque Country* (FBC), called Iparralde in Basque. It will also enable us to compare the somewhat different patterns of agency and choice between men and women among bilingual speakers in the Basque country, as a whole and in the two regions. Main scores showing the basic asymmetry are written in bold fonts in Table 7.7.

Language Competition Models

Table 7.7 Language in use in the family (Basque speakers only) according to territory and gender in Euskal Herria, that is, the Basque country as a whole, 2011. All values in % (apart from absolute values explicitly indicated). Abbreviations: CAB = Basque Autonomous Community; FBC = French Basque Country; tot. = Total, F = Females, M = Males.

	Basque Country (all)			CAB			FBC		
	tot.	M	F	tot.	M	F	tot.	M	F
TOTAL	*100*	*100*	*100*	*100*	*100*	*100*	*100*	*100*	*100*
Sample size (absolute values)	2 628	1 335	1 293	1 689	853	836	320	174	146
Speaking to one's mother									
Most often in Basque	39.7	41.3	38.2	39.7	**41.6**	38.0	46.9	**49.6**	44.0
Both in Basque and Spanish/French	4.2	4.1	4.4	4.0	3.7	4.2	11.2	11.2	11.1
Mainly in Spanish/French (erdara)	56.1	54.6	57.4	56.3	54.7	57.8	41.9	39.2	44.9
Sample size (absolute values)	2 143	2 081	1 062	1 383	693	690	257	142	115
Speaking to one's father									
Most often in Basque	37.3	38.8	35.8	37.5	**39.4**	35.8	43.5	**49.0**	37.4
Both in Basque and Spanish/French	2.7	2.3	3.1	2.3	1.8	2.7	11.1	13.2	8.7
Mainly in Spanish/French (erdara)	60.0	58.9	61.1	60.2	58.9	61.4	45.4	37.8	53.9
Sample size (absolute values)	2 130	1 022	1 108	1 313	621	692	334	179	155
Speaking to one's partner									
Most often in Basque	41.5	44.1	39.4	42.9	**46.3**	40.2	33.3	30.2	**36.3**
Both in Basque and Spanish/French	7.2	7.4	7.0	6.8	6.8	6.8	12.3	14.5	10.2
Mainly in Spanish/French (erdara)	51.3	48.5	53.6	50.3	46.9	53.0	54.4	55.3	53.6
Sample size (absolute values)	1 944	853	1 088	1 179	509	670	328	162	166
Speaking to one's children									
Most often in Basque	70.2	74.2	67.4	74.9	**79.2**	72.0	33.4	**40.6**	28.0
Both in Basque and Spanish/French	11.3	9.2	12.8	10.8	8.0	12.7	14.4	16.0	13.2
Mainly in Spanish/French (erdara)	18.5	16.5	19.8	14.3	12.7	15.3	52.2	43.4	58.8
Sample size (absolute values)	3 447	1 709	1 738	2 190	071	1 119	455	248	207
Speaking at home									
Most often in Basque	39.6	40.3	38.9	40.5	**41.5**	39.6	32.7	31.8	**33.7**
Both in Basque and Spanish/French	10.9	9.5	12.2	10.7	8.6	12.5	15.6	17.4	13.8
Mainly in Spanish/French (erdara)	49.5	50.2	48.9	48.8	49.8	48.0	51.7	50.9	52.5

The first salient result confirms the contrasting models of vernacularity in the FBC versus more vehicularity in the CAB: values are higher for speaking to elders (mother and father: 46.9% with mother in the FBC versus 39.7% in the CAB; 43.5% versus 37.5% when addressing their father) in the former than in the latter, whereas difference becomes drastic as soon as one enters domains of sociability and education. Nevertheless, the figures we just mentioned

Table 7.8 Language used with elders in the CAB and the French Basque Country, 2011.

	Trend	CAB total	FBC total
Speaking to one's mother			
Most often in Basque	ZX	39.7	46.9
Both in Basque and in Spanish/French	YZ	4.0	11.2
Mainly in Spanish/French	ZY	56.3	41.9
Speaking to one's father			
Most often in Basque	ZX	37.5	43.5
Both in Basque and in Spanish/French	YZ	2.3	11.1
Mainly in Spanish/French	ZY	60.2	45.4

match the answer (or the declarative test) for the use 'most often in Basque', which could be represented as a ZX trend. Two other micro-models should be taken into account in interpreting these data: 'Both in Basque and in Spanish/French' on the one hand and 'Mainly in Spanish/French' on the other hand, which could be described as KYZ and KZY (see Table 7.7).

In the next fragment of data (Table 7.9), contrasting patterns of attitudes toward the future and functional worthiness of X clearly oppose the two regions: the ZX model prevails in the CAB roughly 75%, that is, three-quarters of the bilingual population, whereas it gets a rather low score in the FBC (40.6%). The discrepancy between men and women is far stronger in the FBC than in the CAB, though as a general trend, women seem more reluctant to foster Basque in their children's repertoire—but this contrast in gender is much smaller in the CAB: 72% versus 79.2% as opposed to 28% versus 40.6% in the FBC. Nonetheless, cues from the YZ model give fine-grained hints at the motivation of this asymmetric behavior, both in terms of regional and gender variation. Indeed, scores in the YZ cells suggest that mixed usage (X and/or Y) compensate the main bilingual competition model implemented at home with children: parents from the FBC invest more in this pattern of communication than parents from the CAB (14.4% versus 10.8%), and women differ less than men in following this model with their children (13.2% versus 16% in the FBC). When summed up, both models ZX + YZ reach 47.8% in the FBC, and especially 56.6% for men in this region (versus 41% for women, against a 28% score for the ZX model). In the CAB, the YZ strategy added to the ZX reaches a peak of 85.7%, leaving a residual 14.3% to the ZY strategy—against a striking 52.2% in the FBC. These results point at the strong impact of language planning and appreciation (heightened status for X and/or Z) in society. It also points at fine-grained patterns of accommodation of bilingualism in the region deprived of active and efficient language planning, that is, the FBC. In this situation, parents tend to bet more on mixed linguistic behavior, and a modality of Z already observed earlier, which eventually ends up in a lowering of the Z curve, and a strong rise of the Y curve.

Table 7.9 Language used with children in the CAB and the French Basque Country, 2011.

Speaking to one's children	Trend	CAB			FBC		
		Total	Males	Females	Total	Males	Females
Most often in Basque	ZX	74.9	79.2	72.0	33.4	40.6	28.0
Both in Basque & Spanish/French	YZ	10.8	8.0	12.7	14.4	16.0	13.2
Mainly in Spanish/French	ZY	14.3	12.7	15.3	52.2	43.4	58.8

Yet, one should not forget that the situation in both regions differs conspicuously in density of speakers: although the absolute number of Basque speakers is far higher in the CAB than in the FBC, the former are drowned in a huge population of monolinguals, whereas the latter make up a smaller and denser community. The CAB is a wealthy and highly urbanized and industrialized region in northwestern Spain, in a strategic corner of the south-Atlantic economic zone, with two major regional cities (Bilbao/Bilbo and San Sebastian/Donostia), whereas the FBC is mostly a marginal and rural tiny territory, far from regional capitals such as Bordeaux or Toulouse. These contrasting patterns explain why discrepancy in competition models, that is, ZX, YZ, ZY, is far smoother between the two areas, when considering language spoken at home: 40.5% in the CAB versus 32.7% in the FBC for Basque preferentially (see Table 7.7). Here, density of networks in the FBC compensate much of the trend to assimilation, evolving for many decades. Yet, choices are still more clear-cut in the CAB than in the FBC, as only 10.7% of speakers indulge in the YZ model of mixed use, against 15.6% in the FBC. Nevertheless, when summed up, both models still give a score of 51.2% to the CAB. As opposed to 49.2% in the FBC. What might seem at first sight as a mediocre result in the case of the CAB is actually an impressive advance, as compared to the meager result of FBC. Considering the density of its network, this results points at a trend to attrition and decay of the bilingual network in this region. Last, but not least, patterns of use between men and women also point at asymmetric trends: in the FBC, females speak slightly more Basque at home than men (33.7% versus 32.7%). Whereas in the CAB, the asymmetry is reversed: 39.6% versus 41.5%. These figures converge with a model of diglossy and vernacularity in the former case, as opposed to a model of vehicularity and extended function in the latter. It is interesting that the YZ model (i.e., mixed usage) undergoes a striking discrepancy between men in the CAB (8.6%) versus FBC (17.4%), confirming the trend of X attrition in the social space: not only in terms of social space and domains of practice, but also quantitatively, in proportion to utterances and interactions in Y language. Thus, micro-models have been reinterpreted in Table 7.10, on the left column for FBC data (in bold), according to contrastive or converging patterns as ZX > ZY, YZ > YZ, ZY < > Y.

7.7.2 Language use in Catalunya

The next set of data enables us to observe a situation of additive bilingualism from the standpoint of multilingualism as well (languages of migrants and foreign residents and regional minorities, as Aranés, are also taken into account): Catalan in the Generalitat de Catalunya,

Table 7.10 Language used with children in the CAB and the FBC, 2011.

Speaking at home		CAB				FBC		
		Total	Males	Females		Total	Males	Females
Most often in Basque	ZX	40.5	41.5	39.6	ZY	32.7	31.8	33.7
Both in Basque & Spanish/French	YZ	10.8	8.6	12.5	YZ	15.6	17.4	13.8
Mainly in Spanish/French	ZY	48.8	49.8	48.0	Y	51.7	43.4	52.5

Spain. The region is a good example of a structurally open form of bilingualism, as Catalan stands in a structural continuum with Spanish within the domain of Romance languages. Its cost in terms of language acquisition as L2 is far lower than in the case of Euskara as an isolate in relation to Romance languages Spanish and French. As compared to Basque, standard Catalan is fairly easy to learn for Romance speaking citizens and migrants. Therefore, both XZ and YZ trends can be spotted in Table 7.11. Other languages than X and Y will be represented as W; we will give a specific index to W according to whether the language is alloglottic (W_A) or Romance (W_R). The main interesting result here as to the incidence of additive policies is the asymmetry between first language (L_1) and language of identification (i.e., identity and attitudinal adherence to L, which we will label L_d), in comparison with habitual language, that is, the language most functional in everyday life and the work sphere (L_f). Here, for X, the (L_d) function dominates all other, and (L_1) weights even less than (L_f). This hierarchy $L_d \gg L_f \gg L_1$ points at a strong identitary weight for X, strengthened by actual functional worthiness (as L_f prevails on L_1, with 36.29% versus 31.02%, entailing an emerging cohort of speakers with a YZ life trajectory). In the case of Y, the L_1 parameter is the strongest (55.14%), but the lower score of the functional parameter L_f at 50.73%, hints at a YZ trend too, with speakers of Spanish committed to speak in another language in everyday life in Catalunya, although Y is indeed the national language. The discrepancy with the L_d score (47.55%) also suggests commitment to other patterns of identity—among which Catalan and bilingualism. The set 'Both Languages' matches Z, and confirms the previous trends of adherence to regional identity and to bilingualism: 2.44% declare themselves as bilinguals, with a strong identity as such (7%), and current use of both languages (6.82%).

Although the set of data in Tables 7.11 and 7.12 displays low figures, it is nonetheless rewarding to indulge in a careful analysis of results, especially to compare W_R scores to W_A values. Small trends may often provide very valuable hints on the fabric of general configurations as suggested by Penn and Zalesne (2007), and so it seems to be the case here. Tables 7.11 and 7.12 therefore focus on major and minor trends of language use and identification in Catalunya, Spain, according to the official census in 2013. The overall picture is that most of the patterns of multilingualism other than XZ or YZ—that is, the W patterns—seem fairly recent, as they tend to have a scalar decreasing profile from L_1 to L_f, strongly asymmetric to the Catalan and Spanish bilingualism—rather recent too, as it started anew since the 1980 decade, and so did Basque, as mentioned earlier.

Table 7.11 Population aged 15 years and over according to first, identification, and habitual language in Catalunya, Spain, 2013. Source: http://www.idescat.cat/economia/inec?tc=3&id=da01&lang=en. See Table 7.12 for the absolute numbers of the population interviewed in this census, as these figures may especially be relevant for conclusions on small trends of W.

		First language (L_1)	Language of identification (L_d)	Habitual language (L_f)
Catalan	X	31.02	36.38	36.29
Spanish	Y	55.14	47.55	50.73
Both languages	Z	2.44	7.00	6.82
Arabic	W_A	2.43	2.11	1.26
Aranese	W_R	0.04	0.04	0.02
Romanian	W_R	0.90	0.71	0.39
Galician	W_R	0.53	0.20	0.04
Berber	W_A	0.67	0.53	0.39
French	W_R	0.62	0.44	0.20
Portuguese	W_R	0.42	0.31	0.10
Russian	W_A	0.51	0.45	0.22
English	W_A	0.42	0.43	0.42
Italian	W_R	0.47	0.38	0.07
Other languages	$W_?$	3.09	2.45	1.55
Other language combination	$W_?$	0.50	0.54	1.18
Not stated		0.80	0.48	0.32
Total		**100.00**	**100.00**	**100.00**

First, the Z model (i.e., Catalan and Spanish) emerges as the most successful transition. It is nearly the only one to show such a drastic discrepancy between the L_1 parameter and other functions, namely L_d and L_f, and it is unique too in clustering strongly both symbolic and functional roles. Its next competitor in critical mass and in a robust hierachization of functions is Arabic, with decreasing values from L_1 to L_f, since using Arabic or adhering to it as a W language implies having acquired it as a mother tongue (an implication mostly conditioned by its strong discontinuity with Romance languages). Nevertheless, some WR also follow the same pattern as Arabic: Romanian—with similar proportions, French, and Portuguese. On the other hand, Galician and Italian have different profiles, with very low local functionality due to low status among migrants for the former and to the lack of a strong migrant community for the latter. Second, some WA languages also follow similar patterns as Arabic, with even stronger functionality inside the cohort: Berber especially, and Russian to a minor extend as far as L_f is concerned. A third cluster of situations can be identified, though in a very asymmetric way in terms of critical mass: W languages with a very low hierachization of L_1, L_d, and

Table 7.12 Sample of population aged 15 years and over according to first, identification, and habitual language in Catalunya, Spain, 2013.

(Values expressed in thousands of speakers)	First language	Language of identification	Habitual language
Catalan	1940	2275	2270
Spanish	3449	2974	3173
Both languages	153	438	427
Arabic	152	132	79
Aranese	2	2	1
Romanian	56	45	24
Galician	33	13	2
Berber	42	33	24
French	39	27	12
Portuguese	26	19	6
Russian	32	28	14
English	27	27	26
Italian	29	24	4
Other languages	193	153	97
Other language combination	31	34	74
Not stated	50	30	20

L_f parameters: English, with all scores ranking rather high, but flat, and Aranés, with all scores ranking close to zero. Both cohorts actually behave as context-free aggregates, without peculiar problems of status: native speakers of English may use it if solicited to do so, and are probably using it as learning a global language under strong demand, as L_f matches the other two parameters, whereas speakers of Aranés enjoy official status of this Occitan variety in their own small autonomous entity, where they may use it alongside with Spanish and Catalan, but will never be solicited to do so elsewhere in the Generalitat de Catalunya. The evenness of scores, or plateau profile of the figures therefore point at a kind of default aggregate as to criteria of language competition hierarchization.

Conclusion

In the present monograph, we have illustrated a set of tools and concepts related to complex systems that are employed in language dynamics. In order to see how they work and how they relate to linguistics, such tools and concepts have been applied to specific case studies as well as to more general and abstract simplified models. This application follows the general point of view of complex systems theory, that there are some universal frameworks that can be applied through different fields to describe very different types of complex systems.

Language dynamics is a young field looking for an actual integration among the various tools and methods from complex systems, on the one hand, and linguistics, on the other hand, resulting in a unified and consistent picture of language change and use. Such a picture is multifold, in that it concerns (a) the study of the reconstruction of a consistent history of languages, (b) the investigations interpreting the complexity of the currently observed linguistic landscape, and (c) the forecast of the future evolution of language groups. The latter point has recently become of interest due to the fact that cultural and linguistic diversity are now recognized as an invaluable heritage. In fact, the tools of language dynamics have already been used by various authors to suggest language policies and evaluate the ability of a language to find or create the conditions and a suitable niche, in which to be successful.

The structure and contents of this book reflect the current situation: an integration of the main methods of investigation has began, but it is still underway. The path toward the goals of language dynamics is long and necessarily passes through a series of applications to real life situations and historical tests, which are represented by the many linguistic databases now available, containing data about the history of languages and the current sociolinguistic structures. We have not mentioned, for reasons of space, all the databases coming from social networks or from online contents, which offer 'big data' with an unprecedented amount of information and a unique level of fine-grained detail, which are by now a relevant element in linguistic analysis—see for example, Gonçalves and Sánchez (2014); Donoso and Sánchez (2017); Paradowski (2010); Paradowski et al. (2012).

Integration of models and methods will mean especially the integration of the following elements:

1. the different time scales, from the longer one of language origin and evolution to that of every-day language use in the various environments;

2. the spatial scales, from the large scale of the historical population movements accompanied by the spreading of a language to the intricate networks of the social space, possibly coexisting in the same spatial environments, such as urban centers.

The social dimension has become essential: the 'spatial' coordinates used in many language dynamics models may be given different interpretations in terms of age, income classes, social environments, etc., thus allowing the modeler to access many other sides of language dynamics.

So far, we have explored a minimal part of the possible applications of language dynamics. In particular, here we have not analyzed in detail the cognitive side of language, a field that is undergoing fast developments and is becoming a most relevant research line, both for the general understanding of language and for technological applications.

Bibliography

Abrams, D. M., and S. H. Strogatz. 2003. 'Modelling the Dynamics of Language Death'. *Nature* 424: 900.

Albert, R., and A-L. Barabasi, 2002. 'Statistical Mechanics of Complex Networks'. *Review of Modern Physics* 74: 47.

Alinei, M. 2006. 'Darwinism, Traditional Linguistics and the New Palaeolithic Continuity Theory of Language Evolution'. *Evolutionary Epistemology, Language and Culture: A Non-Adaptationist, Systems Theoretical Approach,* edited by Nathalie Gontier, Jean Paul Van Bendegem, and Diederik Aerts, 121–147. Dordrecht: Springer Netherlands.

Amano, T. et al. 2014. 'Global Distribution and Drivers of Language Extinction Risk'. *Proceedings of the Royal Society B: Biological Sciences* 281 (1793): 20141574.

Anderson, P. W. 1972. 'More Is Different'. *Science* 177 (4047): 393–396.

Apostolico, A., and Z. Galil. 1997. *Pattern Matching Algorithms.* Oxford University Press.

Aracil, L. 1982. 'El Bilinguisme Com a Mite'. *Papers de Sociolinguistica, 393–57.* Barcelona: La Magrana. (in Catalan).

Axelrod, R. 1997. 'The Dissemination of Culture—A Model with Local Convergence and Global Polarization'. *Journal of Conflict Resolution* 41 (2): 203–226.

Axelsen, J. B., and S. Manrubia. 2014. 'River Density and Landscape Roughness are Universal Determinants of Linguistic Diversity'. *Proceedings of the Royal Society B: Biological Sciences* 281 (1788): 20141179.

Baggs, I., and H. I. Freedman. 1990. 'A Mathematical Model for the Dynamical Interactions between a Unilingual and Bilingual Population: Persistance versus Extinction'. *Journal of Mathematical Sociology* 16: 51.

———. 1993. 'Can the Speakers of a Dominated Language Survive as Unilinguals—A Mathematical-model of Bilingualism'. *Mathematical and Computer Modelling* 18 (6): 9–18.

Baggs, I., H. I. Freedman, and W. G. Aiello. 1990. 'Equilibrium Characteristics in Models of Unilingual-Bilingual Population Interactions'. *Ocean Waves Mechanics, Computational Fluid Dynamics, and Mathematical Modelling: Proceedings of the 11th International Annual Conference of the Canadian Applied Mathematics Society,* edited by M. Rahman. Southampton: Computational Mechanics Pub. Ltd.

Bakalis, E., and A. Galani. 2012. 'Modeling Language Evolution: Aromanian, an Endangered Language in Greece'. *Physica A: Statistical Mechanics and its Applications* 391 (20): 4963–4969.

Barabási, A-L, and Z. N. Oltvai, 2004. 'Network Biology: Understanding the Cell's Functional Organization. *Nature Reviews* 5: 101–113.

Baronchelli, A. 2016. 'A Gentle Introduction to the Minimal Naming Game'. *Belgian Journal of Linguistics* 30 (1): 171–192.

———. 2018. 'The Emergence of Consensus: A Primer'. *Royal Society Open Science* 5: 172189.

Baronchelli, A., L. Dall'Asta, A. Barrat, and V. Loreto. 2006b. 'Topology-induced Coarsening in Language Games'. *Physical Review EA* 73 (1): 015102.

———. 2007. 'The Role of Topology on the Dynamics of the Naming Game'. *European Physics Journal Special. Topics* 143: 233–235.

Baronchelli, A., M. Felici, V. Loreto, E. Caglioti, and Luc Steels. 2006a. 'Sharp Transition towards Shared Vocabularies in Multi-agent Systems'. *Journal of Statistical Mechanics: Theory and Experiment* 2006 (6): P06014.

Baronchelli, A., V. Loreto, and L. Steels. 2008. 'In-depth Analysis of the Naming Game Dynamics: The Homogeneous Mixing Case'. *International Journal of Modern Physics. C* 19 (5): 785–812.

Beijering, K., C. Gooskens, and W. Heeringa. 2008. 'Predicting Intelligibility and Perceived Linguistic Distance by Means of the Levenshtein Algorithm'. *Linguistics in the Netherlands* 25 (1): 13–24.

Berruto, C. 1989. 'On the Typology of Linguistic Repertoires'. In *Status and Functions of Languages and Language Varieties* edited by U. Ammon. Berlin & New York: de Gruyter.

———. 2004. *Prima lezione di sociolinguistica.* Roma & Bari: Laterza.

Bettinger, R., and M. Baumhoff. 1982. 'The Numic Spread: Great Basin Cultures in Competition'. *American Antiquity* 47 (3): 485–503.

Bickerton, D. 1981. *Roots of Language.* Ann Arbor: Karoma.

———. 1992. *Language and Species.* Chicago: University of Chicago Press.

Binnick, R. I. 1987. 'On the Classification of the Mongolian Languages'. *Central Asiatic Journal* 31 (3/4): 178–195.

Blevins, J. 2004. *Evolutionary Phonology: The Emergence of Sound Patterns.* Cambridge: Cambridge University Press.

Bloomfield, L. 1926. 'A Set of Postulates for the Science of Language'. *Language* 2 (3): 153–164.

Blythe, R. A. 2015. 'Colloquium: Hierarchy of Scales in Language Dynamics'. *The European Physical Journal B* 88 (11): 295.

Bolognesi, R., and W. Heeringa. 2002. 'De Invloed van Dominante Talen op het Lexicon en de Fonologie van Sardische Dialecten'. *Gramma/TTT* 9 (1): 45–84.

Borg, I., and P. Groenen. 2005. *Modern Multidimensional Scaling: Theory and Applications.* New York: Springer.

Box, G. E. P. 1976. 'Science and Statistics'. *Journal of the American Statistical Association* 71 (356): 791–799.

Brown, C. H., Eric W. Holman, Søren Wichmann, and Viveka Velupilla. 2008. 'Automated Classification of the World's Languages: A Description of the Method and Preliminary Results. *STUF, Berlin* 61 (4): 285–308.

Brown, C. H., Eric W. Holman, and Søren Wichmann. 2013. 'Sound Correspondences in the World's Languages'. *Language* 89 (1): 4–29.

Brun-Trigaud, Guylaine, Jean Le Dû, and Yves Le Berre. 2005. *Lectures de l'Atlas Linguistique de la France de J. Gilleron et E. Edmont. Du temps dans l'espace.* CTHS.

Brun-Trigaud, Guylaine, Darlu Pierre, Gaillard-Corvaglia Antonella, Léonard Jean Léo, and Sauzet Patric. 2011. 'Exploration Cladistique de l'ALLOc. In *10éme CIEO (Congrés de l'Associacion Iternacionala d'Estudis Occitans,* edited by : C. Alen Garabato, C.Torreilles, and M. J. Verny. Béziers, France: Lambert-Lucas.

Burridge, J. 2017. 'Spatial Evolution of Human Dialects'. *Physics Review X* 7 (Jul): 031008.

Bybee, J. L., and P. J. Hopper. 2001. *Frequency and the Emergence of Linguistic Structure.* Typological Studies in Language. John Benjamins Publishing Company.

Campbell, L. 1987. 'Tzeltal Dialects: New and Old'. *Anthropological Linguistics* 29 (4): 549–570.

Capen, C. J. 1996. *Diccionario mazateco de Chiquihuitlán.* Tucson: Instituto Lingüístico de Verano.

Capitán, J. A., and S. Manrubia. 2014. *The Ecology of Human Linguistic Groups, 255.* Cd. de México: EditoraC3 CopIt-arXives.

Carro, A., R. Toral, and M. San. 2016. 'Coupled Dynamics of Node and Link States in Complex Networks: A Model for Language Competition'. *New Journal of Physics* 18 (11): 113056.

Casad, E. H., and T. L. Willett, eds. 2000. *Uto-Aztecan: Structural, Temporal, and Geographic Perspectives: Papers in Memory of Wick R. Miller by the Friends of Uto-Aztecan.* Hermosillo, Sonora, México: Editorial Unison, División de Humanidades y Bellas Artes, Universidad de Sonora.

Castellano, C., S. Fortunato, and V. Loreto. 2009. 'Statistical Physics of Social Dynamics'. *Review Modern Physics* 81: 591.

Castelló, X., et al. 2007. 'Anomalous Lifetime Distributions and Topological Traps in Ordering Dynamics'. *Europhysics Letters* 79: 66006.

Castelló, X., A. Baronchelli, and V. Loreto. 2009. 'Consensus and Ordering in Language Dynamics'. *European Physics Journal B* 71: 557.

Castelló, X., et al. 2011. 'Viability and Resilience in the Dynamics of Language Competition'. In *Viability and Resilience of Complex Systems: Concepts, Methods and Case Studies from Ecology and Society,* edited by G. Deuant, and N. Gilbert, 39–74 Springer.

Castelló, X., V. M. Eguíluz, and M. S. Miguel. 2006. 'Ordering Dynamics with Two Non-excluding Options: Bilingualism in Language Competition'. *New Journal Physics* 8 (12): 306.

Cataldo, D. M., M. A. Bamberg, and V. Lucio. 2018. 'Speech, Stone Tool-making and the Evolution of Language'. *PLOS ONE* 13 (1): 1–10.

Cedergren, H. C. J. 1973. 'The Interplay of Social and Linguistic Factors in Panama'. PhD Diss., Cornell University.

Chandler, Daniel. 1994. 'Semiotics for Beginners'. Visual-memory.co.uk/daniel/Documents/S4B/.

———. 2007. *Semiotics: The Basics* 2 edn. London: Routledge.

Chen, G., and Y. Lou. 2010. *Naming Game: Models, Simulations and Analysis (Emergence, Complexity and Computation)*. Switzerland: Springer.

Chomsky, N. 1981. *Lectures on Government and Binding: The Pisa Lectures (Studies in Generative Grammar)* Vol. 9. Dordrecht & Cinnaminson (NJ): Foris.

———. 1986. *Knowledge of Language. Its Nature, Origin, and Use.* New York: Praeger.

———. 1990. *Some Concepts and Consequences of the Theory of Government and Binding.* New York: M.I.T. Press.

Christiansen, M. H., and S. Kirby, eds. 2003. *Language Evolution.* Oxford: Oxford University Press.

Clark, A., C. Fox, and S. Lappin, eds. 2010. *The Handbook of Computational Linguistics and Natural Language Processing.* Oxford: Blackwell.

Clifford, P., and A. Sudbury. 1973. 'A Model For Spatial Conflict'. *Biometrika* 60 (3): 581–588.

Coelho, M. T. P. et al. 2019. 'Drivers of Geographical Patterns of North American Language Diversity'. *Proceedings of the Royal Society B: Biological Sciences* 286 (1899): 20190242.

Collinge, N. E. 1985. *The Laws of Indo-European.* Amsterdam Studies in the Theory and History of Linguistic Science. Amsterdam: John Benjamins Pub. Co.

Colucci, R., Jorge Mira, J. J. Nieto, and M. V. Otero-Espinar. 2016. 'Non Trivial Coexistence Conditions for a Model of Language Competition Obtained by Bifurcation Theory'. *Acta Applicandae Mathematicae* 146 (1): 187–203.

Corvaglia-Gaillard, A. 2012. *From Cladistics to Linguistics: A Study Applied to Southern Italo-Romance and Salentinian Dialects.* Alessandria: Edizioni dell'Orso.

Cowgill, W. 2000. 'A Search for Universals in In-European Diachronic Morphology'. In *New Approaches to Old Problems: Issues in Romance Historical Linguistics*. Current Issues in Linguistic Theory, Vol. 210. Amsterdam: John Benjamins Pub. Co.

Dalmasso, D., V. dellâĂŹAquila, J. L. Léonard. 2012. *2012 ALNum, Numic Cognate Database (Uto-Nahua, Great Bassin, USA), 298 cognats, 5 languages*. Tech. rept. (See final report of the IUF, 2009-14, MAmP project. http://lpp.in2p3.fr/IMG/pdf/jean_le_o_le_onard_iuf_2889-14-final_report.pdf.)

Darwin, C. 1868. *The Variation of Animals and Plants under Domestication*. London: John Murray.

———. 2004. *Descent of Man and Selection in Relation to Sex*. The Barnes & Noble Library of Essential Reading. New York: Barnes & Noble Books.

———. 2007. *On the Origin of Species: By Means of Natural Selection Or the Preservation of Favored Races in the Struggle for Life*. Cosimo Classics Science. New York: Cosimo Classics.

Dauxois, T. 2008. 'Fermi, Pasta, Ulam, and a Mysterious Lady'. *Physics Today* 61 (1): 55.

Davis, I. 1966. 'Numic Consonantal Correspondences'. *International Journal of American Linguistics* 32: 124–140.

Dawkins, R. 1976. *The Selfish Gene*. Oxford: Oxford University Press.

de Oliveira, P. M. C., D. Stauffer, Søren Wichmann, and S. M. de Oliveira. 2008. 'A Computer Simulation of Language Families'. *Journal of Linguistics* 44 (3): 659–675.

de Oliveira, P. M. C., F. W. S. Lima, A. O. Sousa, S. M. de Oliveira, D. Stauffer, and C. Schulze. 2007. 'Bit-strings and Other Modifications of Viviane Model for Language Competition'. *Physica A: Statistical Mechanics and its Applications* 376: 609–616.

de Oliveira, Viviane M., Paulo R. Campos, Marcelo A. Gomes, and Ing Ren Tsang. 2006a. 'Bounded Fitness Landscapes and the Evolution of the Linguistic Diversity'. *Physica A: Statistical Mechanics and its Applications* 368 (1): 257–261.

de Oliveira, Viviane M., Marcelo A. Gomes, and Ing Ren Tsang. 2006b. 'Theoretical Model for the Evolution of the Linguistic Diversity'. *Physica A: Statistical Mechanics and its Applications* 361 (1): 361–370.

de Saussure, Ferdinand. 1915. *Course in General Linguistics*. Translated, with an introduction and notes by Wade Baskin. New York: McGraw-Hill.

Di Chio, C., and P. Di Chio. 2006. 'Simulation Model for the Evolution of Language with Spatial Topology'. In *The Evolution of Language*. Game Theory and Linguistic Meaning, Vol. 18, edited by Angelo Cangelosi, and Andrew D. M. Smith, and Kenny Smith, 51–58. Singapore: World Scientific.

———. 2007. 'Evolutionary Models of Language'. In *Game Theory and Linguistic Meaning*, edited by Ahki-Viekko Pietarinen. Amsterdam: Elsevier.

———. 2009. 'Evolution of Language with Spatial Topology'. *Interaction Studies* 10 (1): 31–50.

Dixon, R. M. W. 1997. *The Rise and Fall of Languages.* Cambridge: Cambridge University Press.

Djordjević, K. 2004. *Configuration Sociolinguistique, Nationalisme et Politique Linguistique, Le cas de la VoÃČÂŕvodine, Hier et Aujourd'hui.* L'Harmattan.

Donoso, G., and D. Sánchez. 2017. 'Dialectometric Analysis of Language Variation in Twitter'. *Proceedings of the Fourth Workshop on NLP for Similar Languages, Varieties and Dialects (VarDial),* 16–25. Valencia, Spain: Association for Computational Linguistics.

Dorogovtsev, S. N., and J. F. F. Mendes. 2003. *Evolution of Networks: From Biological Nets to the Internet and WWW.* Oxford: Oxford University Press.

Ebeling, W., and I. M. Sokolov. 2005. *Statistical Thermodynamics and Stochastic Theory of Nonequilibrium Processes.* Singapore: World Scientific Publishing.

Erdos, P., and A. Rényi. 1959. 'On Random Graphs'. *Publicationes Mathemaiticae (Debrecen)* 6: 290–297.

———. 1960. 'On the Evolution of Random Graphs'. *Publications of the Mathemitical Institute of Hungary Academy of Sciences* 5: 17.

Faisal, A., D. Stout, Jan Apel, and Bruce Bradley. 2010. 'The Manipulative Complexity of Lower Paleolithic Stone Toolmaking'. *PLOS ONE* 5 (11): 1–11.

Fermi, E., J. R. Pasta, and S. M. Ulam. 1955. 'Studies of Nonlinear Problems'. *Los Alamos Scientific Laboratory Report LA-1940,* May.

Fernádndez-Gracia, J., Krzysztof Suchecki, José J. Ramasco, Maxi San Miguel, and Víctor M. Eguíluz. 2014. 'Is the Voter Model a Model for Voters?' *Physics Review Letters* 112 (Apr): 158701.

Fishman, J. A. 1991. *Reversing Language Shift: Theoretical and Empirical Foundations of Assistance to Threatened Languages.* Cleavedon: Multilingual Matters.

———. 2001. 'Why Is It So Hard to Save a Threatened Language?' In *Can Threatened Languages be Saved?* edited by J. A. Fishman, 1–22. Clevedon: Multilingual Matters.

Fitch, W. T. 2010. *The Evolution of Language.* Cambridge: Cambridge University Press.

Fort, J., and T. Pujol. 2008. 'Progress in Front Propagation Research. *Reports on Progress in Physics* 71 (8): 086001.

François, A. 2012. 'The Fynamics of Linguistic Diversity: Egalitarian Multilingualism and Power Imbalance among Northern Vanuatu Languages'. *International Journal of the Sociology of Language* 214: 85–110.

———. 2014. 'Trees, Waves and Linkages: Models of Language Diversification'. In *The Routledge Handbook of Historical Linguistics,* edited by Claire Bowern and Bethwyn Evans, 161–189. London: Routledge.

Freedman, H. I. 1980. *Deterministic Mathematical Models in Population Ecology.* New York: Marcel Dekker.

Gaillard-Corvaglia, A., J-L. Léonard, and P. Darlu. 2007. 'Testing Cladistics on Dialect Networks and Phyla (Galloromance and Southern Italoromance)'. In *Proceedings* of 9th Meeting of the ACL Special Interest Group in Computational Morphology and Phonology, 23–30. SigMorPhon '07. Stroudsburg, PA, USA: Association for Computational Linguistics.

Giordano, M. 2015. *'Analisi di Modelli Dinamici per lâĂŹevoluzione di Popolazioni Bilingue'*. M.Phil. thesis, Universitá degli Studi di Bari, Bari.

Goebl, Hans. 1981. 'Éléments d'Analyse Dialectométrique (avec application á l'AIS)'. *Revue de Linguistique Romane* 45: 349–420.

———. 1984. *Dialektometrische Studien: Anhand Italoromanischer, Rätoromanischer und Galloromanischer Sprachmaterialien aus AIS und ALF (Beihefte zur Zeitschrift für romanische Philologie)*. Tübingen: Niemeyer.

———. 1998. 'On the Nature of Tension in Dialectal Networks. A Proposal for Interdisciplinary Research'. In *Systems: New Paradigms for the Human Sciences* edited by G. Altmann, and W. A. Koch, 549–571. Berlin: Walter de Gruyter.

———. 2003. 'Regards Dialectométriques sur les Données de lâĂŹatlas Linguistique de la France (ALF): Relations Quantitatives et Structures de Profondeur'. *Estudis Romànics*, 25: 59–96.

Goebl, Hans, and Guillaume Schiltz. 1997. 'A Dialectometrical Compilation of CLAE 1 and CLAE 2. Isoglosses and Dialect Integration'. In *Computer Developed Linguistic Atlas of England (CLAE)*, Vol. 2, edited by W. Viereck, and H. Ramisch, 13–21. Tübingen: Max Niemeyer Verlag.

Goebl, Hans, Slawomir Sobota, and Edgar Haimerl. 2002. 'Analyse Dialectométrique des Structures de Profondeur de l'ALF'. *Revue de Linguistique Romane* 66: 5–63.

Gomes, M. A. F., G. L. Vasconcelos, I. J. Tsang, and Ing Ren Tsang. 1999. 'Scaling Relations for Diversity of Languages'. *Physica A: Statistical Mechanics and its Applications* 271 (3-4): 489–495.

Gonçalves, B., and D. Sánchez. 2014. 'Crowdsourcing Dialect Characterization through Twitter'. *PLOS ONE* 9 (11): 1–6.

Government of Oaxaca. 2016. *Microregión 13: Zona Mazateca*. Oaxaca. Secretaria de Desarrollo Social y Humano http://www.coplade.oaxaca.gob.mx/wp-content/uploads/2011/09/Microrregion13.pdf.

Grayson, D. K. 2011. *The Great Basin: A Natural Prehistory*. Berkeley: University of California Press.

Greenberg, J. H. 1960. 'A Quantitative Approach to the Morphological Typology of Language'. *International Journal of American Linguistics* 26 (3): 178–194.

Grishman, R. 1986. *Computational Linguistics: An Introduction*. Cambridge: Cambridge University Press.

Gudschinsky, S. C. 1955. 'Lexico-Statistical Skewing from Dialect Borrowing'. *International Journal of American Linguistics* 21 (2): 138–149.

———. 1958. 'Mazatec Dialect History'. *Language* 34: 469–481.

———. 1959. 'Proto-Popotecan. A Comparative Study of Popolocan and Mixtecan'. *Supplementary to International Journal of American Linguistics, Memoir 15*, 25 (2).

Harney, H. L. 2003. *Bayesian Inference: Data Evaluation and Decision*. New York: Springer.

Hastie, T., R. Tibshirani, and J. Friedman. 2009. *The Elements of Statistical Learning. Data Mining, Inference, and Prediction*. 2nd edn. Springer Series in Statistics. New York: Springer-Verlag.

Hastings, A. 1996. *Population Biology: Concepts and Models*. New York: Springer.

Heeringa, W., and C. Gooskens. 2003. 'Norwegian Dialects Examined Perceptually and Acoustically'. *Computers and the Humanities*, 37 (3): 293–315.

Heeringa, W. J. 2004. 'Measuring Dialect Pronunciation Differences using Levenshtein Distance'. PhD thesis, University of Groningen.

Heggarty, P. 2017. 'Towards a (Pre)History of Linguistic Convergence Areas: Correlates in Genetics, Archaeology, History and Geography'. In *Mémoires de la Société de Linguistique, Nouvelle Série*, Vol. 23, edited by J. L. Léonard, 135–178.

Heinsalu, Els, Marco Patriarca, and Jean Léo Léonard. 2014. 'The Role of Bilinguals in Language Competition'. *Advances in Complex Systems*, 17 (1): 1450003.

Hickey, R., ed. 2010. *The Handbook of Language Contact*. Hoboken, New Jersey: Wiley-Blackwell.

Holley, R., and T. Liggett,. 1975. 'Ergodic Theorems for Weakly Interacting Systems and the Voter Model'. *Annals of Probability* 3: 643.

Holman, Eric W., Søren Wichmann, Cecil H. Brown, and E. Anthon Eff. 2015. 'Diffusion and Inheritance of Language and Culture: A Comparative Perspective'. *Social Evolution & History*, 14 (1): 49–64.

Hopkins, N. 1965. 'Great Basin Prehistory and Uto-Aztecan'. *American Antiquity* 31 (1): 48–60.

Hruschka, D. J. et al. 2009. 'Building Social Cognitive Models of Language Change. *Trends in Cognitive Sciences* 13 (11): 464–469.

Hurford, J. R. 1989. 'Biological Evolution of the Saussurean Sign as a Component of the Language-acquisition Device'. *Lingua* 77: 187–222.

———. 1994. 'Linguistics and Evolution: A Background Briefing for Non-Linguists'. *Current Opinion Neurobiolgy* 10 (1): 149–157.

———. 2014. *Origins of Language: A Slim Guide*. Oxford: Oxford University Press.

Hyman, L. 2011. 'The Macro-Sudan Belt and Niger-Congo Reconstruction'. *Language Dynamics and Change* 1 (1): 3–49.

Iannucci, D. 1972. 'Numic Historical Phonology'. PhD thesis, Cornell University.

Iriberri, N., and J-R. Uriarte. 2012. 'Minority Language and the Stability of Bilingual Equilibria'. *Rationality and Society* 24 (4): 442–462.

Isern, N., and J. Fort. 2014. 'Language Extinction and Linguistic Fronts'. *Journal of The Royal Society Interface* 11 (94): 20140028.

Isern Sardó, 2011. 'Front Spreading in Population Dynamic Models. Theory and Application to the Neolithic Transition'. PhD thesis, Universitst de Girona, Girona.

Jackendoff, R. 1985. *Semantics and Cognition.* Vol. 32. Massachusetts Institute of Technology.

———. 1990. *Semantic Structures.* Vol. 32. Massachusetts: MIT Press.

———. 1997. *The Architecture of the Language Faculty.* Linguistic Inquiry Monograph, Vol. 32. Massachusetts: MIT Press.

Jamieson, C. C., K. Voigtlander, and A. R. Jamieson. 1988. *Gramática Mazateca del Municipio de Chiquihuitlán, Oaxaca.* México, D. F.: Instituto Lingüístico de Verano.

Kandler, A. 2009. 'Demography and Language Competition'. *Human Biology* 81: 181.

Kandler, A., and J. Steele. 2008. 'Ecological Models of Language Competition'. *Biological Theory* 3: 164.

Kandler, A., R. Unger, and J. Steele. 2010. 'Language Shift, Bilingualism and the Future of Britain's Celtic Languages'. *Philosophical Transactions of the R. Society B* 365: 3855.

Karjus, A., and M. Ehala. 2015. *Linguistic Attitudes in Estonia.*

———. 2018. 'Testing an Agent-based Model of Language Choice on Sociolinguistic Survey Data'. *Language Dynamics and Change* 8 (2): 219–252.

Karklins, R. 2000. 'Theories of National Integration and Developments in Latvia'. In *Integrācija un etnopolitika*, edited by E. Vēbers. Riga: Latvijas Universitātes, Filosofijas un Sociologijas Institūts.

Kaufman, T., and J. Justeson. 2003. *A Preliminary Mayan Etymological Dictionary.* Foundation for the Advancement of Mesoamerican Studies.

Keresztes, L. 1999. *Development of Mordvin Definite Conjugation.* Suomalais-Ugrilainen Seura: Suomalais-Ugrilaisen Seuran toimituksia, Vol. 233. Finno-Ugrian Societya.

Kihm, A. 2006 (October). 'Le paradoxe créole'. In *Faire sens. Colloque en hommage à Pierre Encrevé.*

———. 2014. 'Mazatec Verb Inflection. Revisiting Pike (1948) and Comparing Four Dialects'. In *Patterns in Mesoamerican Morphology,* edited by J.-L. Léonard, 26–76. Paris: Michel Houdiard Editeur.

Killion, W. T., and J. Urcid. 2001. 'The Olmec Legacy: Cultural Continuity and Change in México's Southern Gulf Coast Lowlands'. *Journal of Field Archeology* 28: 3–25.

Kirk, P. L. 1966. 'Proto-Mazatec Phonology'. PhD thesis, University of Washington, Washington.

———. 1970. 'Dialect Intelligibility Testing: The Mazatec Study'. *International Journal of American Linguistics* 36 (3): 205–211.

Komarova, N. L., and M. A. Nowak. 2001a. 'The Evolutionary Dynamics of the Lexical Matrix'. *Bulletin of Mathematical Biology* 63 (3): 451–484.

———. 2001b. 'Natural Selection of the Critical Period for Language Acquisition'. *Proceedings: Biological Sciences* 268 (1472): 1189–1196.

Kortmann, B., ed. 2004. *Dialectology Meets Typology: Dialect Grammar from a Cross-linguistic Perspective*. Berlin: Mouton de Gruyter.

Kosmidis, K., J. M. Halley, and P. Argyrakis. 2005. 'Language Evolution and Population Dynamics in a System of Two Interacting Species'. *Physica A: Statistical Mechanics and its Applications* 353: 595–612.

Krupa, V. 1966. *Morpheme and Word in Maori*. Vol. 46. London: Mouton & Co.

Labov, W. 1970. *The Study of Non-standard English*. Eric reports. Washington: National Council of Teachers of English.

———. 1972a. *Language in the Inner City: Studies in the Black English Vernacular*. Conduct and Communication. Philadelphia: University of Pennsylvania Press.

———. 1972b. *Sociolinguistic Patterns*. Conduct and Communication. Philadelphia: University of Pennsylvania Press.

———. 2006. *The Social Stratification of English in New York City*. Cambridge: Cambridge University Press.

Lamb, S. 1958. 'Linguistic Prehistory in the Great Basin'. *International Journal of American Linguistics* 24 (2): 95–100.

Lauristin, M., and M. Heidmets, eds. 2002. *The Challenge of the Russian Minority: Emerging Multicultural Democracy in Estonia*. Tartu: Tartu University Press.

Law, Daniel Aaron. 2011. 'Linguistic Inheritance, Social Difference, and the Last Two Thousand Years of Contact Among Lowland Mayan Languages.' PhD thesis, The University of Texas at Austin.

Lenaerts, Tom, Bart Jansen, Karl Tuyls, and Bart De Vylder. 2005. 'The Evolutionary Language Game: An Orthogonal Approach'. *Journal of Theoretical Biology* 235 (4): 566–582.

Léonard, Jean Léo. 2007. 'The Use of Basque in the Light of the Theory of Integration'. *Cross-Cultural Studies Series CILTRANS Occasional Papers* 103. http://jll.smallcodes.com /articles.7/-laquo-Donn-eacuteesstatistiques-de-la-pratique-du- basque-e.page.

———. 2014. 'Le Sprachbund Mésoaméricain: Instanciation Spatiale dâĂŹun Concept Opératoire'. *Faits de Langues* 43 (1): 11–40.

———, ed. 2016. *Actualité des Néogrammairiens*. Mémoires de la Société de Linguistique de Paris, Vol. 23. Paris: EFL - Empirical Foundations of Linguistics.

Léonard, Jean Léo, Vittorio Dell'Aquila, and Antonella Gaillard-Corvaglia. 2012. 'The ALMaz (Atlas Lingüístico Mazateco): From Geolinguistic Data Processing to Typological Traits'. *STUF, Akademie Verlag* 65 (1): 78–94.

Léonard, Jean Léo, and Julien Fulcrand. 2016. 'Tonal Inflection and Dialectal Variation in Mazatec'. In *Tone and Inflection: New Facts and New Perspectives*. Trends in Linguistics. Studies and Monographs [TiLSM], Vol. 269. Berlin: De Gruyter.

Léonard, Jean Léo, M. Patriarca, P. Darlu, and E. Heinsalu. 2015. 'Modeling Regional Variation from EAS: Complexity and Communal Aggregates'. In *Linguistic Variation in the Basque Language and Education-I âĂŤ Euskararen Bariazioa eta Bariazioaren Irakaskuntza-I*, edited by G. Olabarri, A. Aurrekoetxea, A. Romero, and A. E. Lejarreta, 145–172. Euskal Herriko Unibertsitateko Argi-talpen Zerbitzua – Servicio Editorial de la Universidad del País Vasco.

Léonard, Jean Léo, Marco Patriarca, Els Heinsalu, Kiran Sharma, and Anirban Chakraborti. 2017. 'Patterns of Linguistic Diffusion in Space and Time: The Case of Mazatec'. In *Econophysics and Sociophysics: Recent Progress and Future Directions,*. edited by F. Abergel, et al. 227–251, Chapter 17 New Economic Windows. New York: Springer.

———. 2019. 'Patterns of Linguistic Diffusion in Space and Time: The Case of Mazatec'. In *Complexity Applications in Language and Communication Sciences* edited by A. M. Massip-Bonet, G. Bel-Enguix, and A. Bastardas-Boada. New York: Springer.

Léonard, Jean Léo, Clément Vulin, and Pierre Darlu. 2010. 'Application de la Cladistique aux Langues Mésoaméricaines'. *Journée d'études de la Société Française de Systématique*.

Levenshtein, V. 1. 1966. 'Binary Codes Capable of Correcting Deletions, Insertions, and Reversals'. *Soviet Physics Doklady* 10 (8): 707–710.

Lieberman, P. 1984. *The Biology and Evolution ofLanguage*. Harvard University Press.

Liggett, T. M. 1999. *Stochastic Interacting Sysyems: Contact, Voter and Exclusion Processes*. New York: Springer.

Lipowska, D. 2011. 'Naming Game and Computational Modelling of Language Evolution'. *Computational Methods in Science and Technology* 17 (1–2): 41–51.

Lipowski, A., and D. Lipowska. 2008. 'Bio-linguistic Transition and Baldwin Effect in an Evolutionary Naming-game Model'. *International Journey of Modern Physics C* 19 (3): 399–407.

———. 2009. 'Language Structure in the n-object Naming Game'. *Physics Review E* 80 (5): 056107.

———. 2018. 'Emergence of Linguistic Conventions in Multi-agent Reinforcement Learning'. *PLOS ONE* 13 (11): 1–18

Lomolino, M. V., B. R. Riddle, and J. H. Brown. 2006. *Biogeography*. Sunderland, Masachusetts: Sinauer Associates.

Loreto, V., and L. Steels. 2007. 'Emergence of Language'. *Nature Physics,* 3, 758.

Loreto, V., et al. 2011. 'Statistical Physics of Language Dynamics'. *Journal of Statistical Mechanics: Theory and Experiment* 2011 (4): P04006.

Madsen, D. 1975. 'Dating Paiute-Shoshoni Expansion in the Great Basin'. *American Antiquity* 40 (1): 82–86.

Marchetti, G., M. Marco, and E. Heinsalu. 2019. 'A Bird's-Eye View of Naming Game Dynamics', arXiv.org

Toledo, E. M. (B'alam Q'uq). 1999. *La Cuestión Akateko-q'anjob'al, Una Comparación Gramatical*. Guatemala: Universidad Mariano Gálvez de Guatemala, Facultad de Humanidades, Escuela de Lingüística.

McMahon, A., and R. McMahon. 2013. *Evolutionary Linguistics*. Cambridge Textbooks in Linguistics. Cambridge: Cambridge University Press.

Mende, W, and K. Wermke. 2006. 'A Long Way to Understanding Cultural Evolution'. *Behavioral and Brain Sciences*, 29 (4): 358.

Mesthrie, R. 2011. *The Cambridge Handbook of Sociolinguistics*. Cambridge: Cambridge University Press.

Miller, G. A. 1970. *The Psychology of Communication: Seven Essays*. London: Penguin Books.

———. 1981. *Language and Speech*. San Francisco: W. H. Freeman & Co.

Milroy, J., and L. Milroy. 1978. 'Belfast: Change and Variation in an Urban Vernacular'. In *Sociolinguistic Patterns in British English*, edited by P. Trudgill. Pennsylvania: University Park Press.

Milroy, L. 1980. *Language and Social Networks*. Language in Society. Hoboken, New Jersey: Wiley.

Minett, J. W., and W. S-Y. Wang. 2008. 'Modelling Endangered Languages: The Effects of Bilingualism and Social Structure'. *Lingua* 118: 19.

Mira, J., and A. Paredes. 2005. 'Interlinguistic Similarity and Language Death Dynamics'. *Europhysics Letters* 69: 1031.

Mira, J., L. F. Seoane, and J. J. Nieto. 2011. 'The Importance of Interlinguistic Similarity and Stable Bilingualism When two Languages Compete'. *New Journal of Physics* 13: 033007.

Morgan, T. J. H. et al. 2015. 'Experimental Evidence for the Co-evolution of Hominin Tool-making Teaching and Language'. *Nature Communications* 6 (Jan): 6029.

Morpurgo A. D. 1998. *Nineteenth-century Linguistics*. History of linguistics, Vol. 4. London & New York: Routledge.

Mufwene, S. 2001. *The Ecology of Language Evolution*. Cambridge: Cambridge University Press.

———. 2008. *Language Evolution: Contact, Competition and Change*. London: Continuum International Publishing Company.

———. 2013a. 'The Ecology of Language: Some Evolutionary Perspectives'. In *Da Fonologia á Ecolínguistica – Ensaios em homenajem a Hildo Honório do Couto*, edited by E. K. N. N. do Couto, D. B. de Albuquerque, and G. P. de Araújo. Brasilia: Thesaurus.

———. 2013b. 'The Emergence of Complexity in Language: An Evolutionary Perspective'. In *Complexity Perspectives on Language, Communication, and Society*, edited by A. Massip-Bonet and A. Bastardas-Boada. Heidelberg: Springer.

———. 2013c. 'What African Linguistics Can Contribute to Evolutionary Linguistics'. In *Selected Proceedings of the 43th Annual Conference on African Linguistics* edited by O. O. Olanik and K. W. Sanders. Somerville (MA): Cascadilla Proceedings Project.

Murray, J. D. 2002. *Mathematical Biology I: An Introduction.* New York: Springer.

———. 2003. *Mathematical Biology II: Spatial Models and Biomedical Application.* New York: Springer

Nerbonne, J., and W. Kretzschmar. 2003. 'Introducing Computational Techniques in Dialectometry'. *Computers and the Humanities* 37 (3): 245–255.

Nichols, J. 1986. 'Head-Marking and Dependent-Marking Grammar'. *Language* 62 (1): 56–119.

Nie, Lin-Fei, Zhi-Dong Teng, Juan J. Nieto, and Il Hyo Jung. 2013. 'Dynamic Analysis of a Two-Language Competitive Model with Control Strategies'. *Mathematical Problems in Engineering* 654619.

Nowak, Martin A., D. C. Krakauer, and A. Dress. 1999a. 'An Error Limit for the Evolution of Language'. *Proceedings of the Royal Society of London. Series B: Biological Sciences* 266 (1433): 2131–2136.

Nowak, Martin A. 2000. 'Evolutionary Biology of Language'. *Philos. Trans. R. Soc. London B Biol. Sci.,* 355 (1403): 1615–22.

Nowak, Martin A., N. L. Komarova, and P. Niyogi. 2001. 'Evolution of Universal Grammar'. *Science* 291 (5501): 114–118.

———. 2002. 'Computational and Evolutionary Aspects of Language'. *Nature* 417 (6889): 611–617.

Nowak, Martin A., and D. C. Krakauer. 1999. 'The Evolution of Language'. *Proceedings of the National Academy of Sciences,* 96 (14): 8028–8033.

Nowak, Martin A., J. B. Plotkin, and D. C. Krakauer. 1999b. 'The Evolutionary Language Game'. *Journal of Theoretical Biology* 200 (2): 147–162.

Oliphant, M., and J. Batali. 1996. 'Learning and the Emergence of Coordinated Communication'. *The Newsletter of the Center for Research on Language* 11 (1): 146.

Ostler, N. 2005. *Empires of the Word: A Language History of the World.* London: Harper Collins.

Otero-Espinar, M. V., Luis F. Seoane, Juan J. Nieto, and Jorge Mira. 2013. 'An Analytic Solution of a Model of Language Competition with Bilingualism and Interlinguistic Similarity'. *Physica D: Nonlinear Phenomena* 264: 17–26.

Owolabi, K. M., and J. F. Gómez-Aguilar. 2018. 'Numerical Simulations of Multilingual Competition Dynamics with Nonlocal Derivative'. *Chaos, Solitons & Fractals* 117: 175–182.

Palau, A. D., ed. 2002. *Una Radiografia Social de la Llengua Catalana.* Generalitat de Catalunya, Departament de Cultura, Cossetánia.

Paradowski, M. B. 2010. 'A Complexity Science Perspective on Language Spread'. In *Proceedings of ISCA Tutorial and Research Workshop on Experimental Linguistics*, edited by A. Botinis, 137–140. Athens: University of Athens.

Paradowski, M. B., Chih-Chun Chen, Agnieszka Cierpich, and Lukasz Jonak. 2012. 'From Linguistic Innovation in Blogs to Language Learning in Adults: What Do Interaction Networks Tell Us?' In *Proceedings of the Conference on 'Social Computing, Social Cognition, Social Networks, and Multiagent Systems'*, edited by G. Dodig-Crnkovic, et al. The Society for the Study of Artificial Intelligence and Simulation of Behaviour (aisb.org.uk).

Pastor-Satorras, Romualdo, Claudio Castellano, Piet Van Mieghem, and Alessandro Vespignani. 2015. 'Epidemic Processes in Complex Networks'. *Review of Modern Physics* 87 (3): 925–979.

Patriarca, M., and E. Heinsalu. 2009. 'Influence of Geography on Language Competition'. *Physica A* 388: 174.

Patriarca, M., and T. Leppänen. 2004. 'Modeling Language Competition'. *Physica A* 338: 296.

Patriarca, Marco, Xavier Castelló, José Ramón Uriarte, Víctor M. Eguíluz, and Maxi San Miguel. 2012. 'Modeling Two-language Competition Dynamics'. *Adv. Comp. Syst.* 15 (3&4): 1250048.

Penn, M., and K. Zalesne. 2007. *Microtrends: Surprising Tales of the Way we Live Today*. London: Penguin.

Pike, K. L. 1948. *Tone Languages: A Technique for Determining the Number and Type of Pitch Contrasts in a Language, with Studies in Tonemic Substitution and Fusion*. Linguistics. Ann Arbor, Michigan: University of Michigan Press.

Pinasco, J. P., and L. Romanelli. 2006. 'Coexistence of Languages is Possible'. *Physical A* 361: 355.

Plotkin, J., and M. Nowak. 2001. 'Major Transitions in Language Evolution'. *Entropy* 3 (4): 227–246.

Polian, G., Jean Léo Léonard, Els Heinsalu, and Marco Patriarca. 2014. 'Variación Dialectal Del Tseltal (Maya Occidental) En Los Ámbitos Morfológico, Fonológico Y Léxico: Un Enfoque Holístico'. In *Patterns In Mesoamerican Morphology* edited by J. L. Léonard and A. Kihm. Michel Houdiard Éditeur.

Poppe, N. 1955. *Introduction to Comparative Mongolian Studies*. Helsinki: Finno-Ugrian Society (SUS – Suomalais-Ugrilainen Seura).

Pucci, L., P. Gravino, and V. D. P. Servedio. 2014. 'Modeling the Emergence of a New Language: Naming Game with Hybridization'. In *Self-Organizing Systems*, edited by W. Elmenreich, F. Dressler, and V. Loreto, 78–89. Berlin, Heidelberg: Springer.

Ramírez, A. G. 1996. 'Dialectología y Sociolinguística'. In *Manual de Dialectología Hispánica: El Español de España* edited by M. Alvar, 37–48. Ariel Letras. Barcelona: Editorial Planeta.

Rannap, J. 2017. *'Mathematical Analysis of Numic Languages'*. Bachelor's thesis, University of Tartu, Tartu.

Rogers, E. M. 1983. *Diffusion of Innovations*. New York: Free Press.

Ross, John, and A. P. Arkin. 2009. 'Complex Systems: From Chemistry to Systems Biology'. *Proceedings of the National Academy of Sciences* 106 (16): 6433–6434.

Rowe, B. M., and D. P. Levine. 2015. *A Concise Introduction to Linguistics*. London: Routledge.

Scheer, T. 2004. *A Lateral Theory of Phonology: What is CVCV, and Why Should It Be?* Vol. 1. Berlin: Mouton de Gruyter.

———. 2011. *A Guide to Morphosyntax-Phonology Interface Theories: How Extra-Phonological Information is Treated in Phonology since Trubetzkoy's Grenzsignale*. Vol. 1. Berlin: Mouton de Gruyter.

———. 2012. *Direct Interface and One-Channel Translation. A Non-Diacritic Theory of the Morphosyntax-Phonology Interface*. Vol. 2. Berlin: Mouton de Gruyter.

Scheer, T., and P. Ségéral. 2016. 'L'actualité des Néogrammairiens'. *Memoires de la Société de Linguistique de Paris* 23: 15–68.

Schelling, T. C. 1978. *Micromotives and Macrobehavior*. New York: W. W. Norton & Co.

Schmidt, J. 2017. *Die Verwandtschaftsverhaltnisse der Indogermanischen Sprachen*. Weimar: H. Bohlau.

Schuchardt, H. E. M. 1980. *Pidgin and Creole Languages, Selected Essays by Hugo Schuchardt* (Translated by Glenn G. Gilbert.). Cambridge: Cambridge University Press.

Schulze, C., and D. Stauffer. 2005. 'Monte Carlo Simulation of the Rise and the Fall of Languages'. *International Journal of Modern Physics C* 16 (05): 781–787.

———. 2006. 'Recent Developments in Computer Simulations of Language Competition'. *Computing in Science & Engineering* 8 (3): 60–67.

———. 2007. 'Competition of Languages in the Presence of a Barrier'. *Physica A: Statistical Mechanics and its Applications* 379 (2): 661–664.

Schulze, C., D. Stauer, and S. Wichrnann. 2008. 'Birth, Survival and Death of Languages by Monte Carlo Simulation'. *Communications in Computational Physics* 3 (2): 271–294.

Schumann, O. 1983. 'La Relación Lingüística Chuj-tojolabal. In *Los Legítimos Hombres. Aproximación Antropológica al Grupo Tojolabal,* Vol. 1, edited by M. H. Ruz. México: Universida Nacional Autónoma de México.

Schwämmle, V. 2005. 'Simulation for Competition of Languages with an Aging Sexual Population'. *International Journal of Modern Physics C* 16 (10): 1519–1526.

Schweitzer, F. 2003. *Brownian Agents and Active Particles*. Berlin: Springer-Verlag.

Shaul, D., and S. Ortman. 2014. *A Prehistory of Western North America: The Impact of Uto-Aztecan Languages*. Albuquerque: University of New México Press.

Shen, Z. 1997. 'Exploring the Dynamic Aspect of Sound Change' (PhD thesis). *Journal of Chinese Linguistics* Monograph Series 11. Hong Kong: Chinese University Press.

Sofuoglu, Y., and N. Ozalp. 2017. 'Fractional Order Bilingualism Model Without Conversion from Dominant Unilingual Group to Bilingual Group'. *Differential Equations and Dynamical Systems* 25 (1): 1–9.

Solé, R. V., B. Corominas-Murtra, and J. Fortuny. 2010. 'Diversity, Competition, Extinction: the Ecophysics of Language Change. *Journal of The Royal Society Interface* 7: 1647–1664.

Sørensen, A. 1967. 'Multilingualism in the Northwest Amazon'. *American Anthropologist* 69: 670–684.

Stauffer, D., and C. Schulze. 2005. 'Microscopic and Macroscopic Simulation of Competition Between Languages'. *Physics Life Reviews* 2: 89.

Stauffer, D., C. Schulze, F. Welington S. Lima, S. Wichmann, and Sorin Solomon. 2006. 'Non-equilibrium and Irreversible Simulation of Competition Among Languages'. *Physica A: Statistical Mechanics and Its Applications* 371 (2): 719–724.

Stauffer, D., X. Castelló, V. M. Eguiluz, and Maxi San Miguel. 2007. 'Microscopic Abrams-Strogatz Model of Language Competition. *Physica A: Statistical Mechanics and its Applications* 374 (2): 835–842.

Stenzel, K. 2005. 'Multilingualism in the Northwest Amazon, Revisited'. In *Memorias del Congreso de Idiomas Indígenas de Latinoamerica,* Vol. II. University of Texas at Austin.

Strogatz, S. H. 1994. *Nonlinear Dynamics and Chaos: With Applications to Physics, Biology, Chemistry and Engineering.* Reading, Massachusetts: Perseus Books.

Sutantawibul, Chosila, Pengcheng Xiao, Sarah Richie, and Daniela Fuentes-Rivero. 2018. 'Revisit Language Modeling Competition and Extinction: A Data-Driven Validation'. *Journal of Applied Mathematics and Physics* 6 (7): 1558–1570.

Sutherland, W. J. 2003. 'Parallel Extinction Risk and Global Distribution of Languages and Species'. *Nature* 423 (May): 276.

Svantesson, J. O. Anna Tsendina, Anastasia Karlsson, and Vivan Franzen. 2005. *The Phonology of Mongolian.* Oxford: Oxford University Press.

Taagepera, R. 2008. *Making Social Sciences More Scientific: The Need for Predictive Models.* Oxford: Oxford University Press.

Tallerman, M., and K. R. Gibson. 2012. *The Oxford Handbook of Language Evolution.* Oxford Handbooks in Linguistics. Oxford: Oxford University Press.

Teşileanu, T., and H. Meyer-Ortmanns. 2006. 'Competition of Languages and their Hamming Distance'. *International Journal of Modern Physics C* 17 (2): 259–278.

Toivonen, Riitta, Xavier Castelló, Víctor M. Eguíluz, Jari Saramäki, K. Kaski, and Maxi San Miguel. 2009. 'Broad Lifetime Distributions for Ordering Dynamics in Complex Networks'. *Physics Review E* 79: 016109.

Tort, P., A. Schleicher, and D. Modigliani. 1980. *Évolutionnisme et Linguistique.* Paris: J. Vrin.

Trapa, P. E., and M. A. Nowak. 2000. 'Nash Equilibria for an Evolutionary Language Game'. *Journal of Mathematical Biology* 41 (2): 172–188.

Tria, Francesca, Vito DP Servedio, Salikoko S. Mufwene, and Vittorio Loreto. 2015. 'Modeling the Emergence of Contact Languages'. *PLOS ONE* 10 (4): 1–11.

Tuite, K. 1988. 'Kartvelian Morphoyntax: Number Agreement and Morphosyntactic Orientation in the South Caucasian Languages'. PhD thesis, Chicago University.

Unamuno, Lorea, Ariane Ensunza, Aitor Iglesias, and Jose Luis Ormaetxea. 2012 (01). 'EAS Project: First Data on Syntactic Variation in the Basque Language. In *Proceedings of the International Symposium on Limits and Areas in Dialectology (LimiAr 2011, Lisbon)* edited by A. Pérez et al., 225–235.

Uriarte, J-R. 2016. *A Game-Theoretic Analysis of Minority Language Use in Multilingual Societies*. London: Palgrave Macmillan. 689–711.

van Kampen, N. G. 1992. *Stochastic Processes in Physics and Chemistry*. Amsterdam: Elsevier.

Vazquez, F., X. Castelld, and M. San Miguel. 2010. 'Agent Based Models of Language Competition: Macroscopic Descriptions and Order-disorder Transitions'. *Journal of Statistical Mechanics* P04007.

Villalobos, F., and T. F. Rangel. 2014. *Geographic Patterns of Biodiversity*. Cd. de Mdxico: EditoraC3 CopIt-arXives. Page 1.

Vogt, P. 2009. 'Modeling Interactions between Language Evolution and Demography'. *Human Biology* 81 (3): 237–258.

Wang, W. S-Y., and J. W. Minett. 2005. 'The Invasion of Language: Emergence, Change and Death'. *TRENDS in Ecology and Evolution* 20: 263.

Watts, D. J., and S. H. Strogatz. 1998. 'Collective Dynamics of "Small-world" Networks. *Nature* 393 (6684): 440.

Wendel, J. 2005. 'Notes on the Ecology of Language'. *Bunkyo Gakuin University Academic Journal* 5: 51–76.

Wichmann, S. 2005. 'On the Power-law Distribution of Language Family Sizes'. *Journal of Linguistics* 41 (1): 117–131.

———. 2008a. 'The Emerging Field of Language Dynamics'. *Language and Linguistics Compass* 2 (3): 442–455.

———. 2008b. 'Teaching & Learning Guide for: The Emerging Field of Language Dynamics'. *Language and Linguistics Compass* 2 (6): 1294–1297.

Wichmann, S., E. W. Holman, D. Bakker, and Cecil H. Brown. 2010. 'Evaluating Linguistic Distance Measures'. *Physica A* 389 (17): 3632–3639.

Wyburn, J., and J. Hayward. 2008. 'The Future of Bilingualism: An Application of the Baggs and Freedman Model'. *Journal of Mathematical Society* 32 (4): 267–284.

———. 2009. 'OR and Language Planning: Modeling the Interaction between Unilingual and Bilingual Populations'. *Journal of Operation Research Society* 60 (5): 626–636.

———. 2010. 'A Model of Language-Group Interaction and Evolution Including Language Acquisition Planning'. *Journal of Mathematical Sociology* 34 (3): 167–200.

Xia, H., Simon J. Greenhill, Marcel Cardillo, Hilde Schneemann, and Lindell Bromham. 2019. 'The Ecological Drivers of Variation in Global Language Diversity'. *Nature Communications* 10 (1): 2047.

Young, D., and R. Bettinger. 1992. 'The Numic Spread: A Computer Simulation'. *American Antiquity* 57 (1): 85–99.

Index

adaptation, 13–14, 61, 98
adaptive system, 11, 13
adjacency matrix, 22–23
artificial intelligence, 11, 101
assimilation, 16–17, 32, 73, 165
automated text analysis, 3

backness, 20
Basque, 4, 41, 57–59, 65, 104, 109, 141, 164, 174
bilingual, 15, 98, 104, 120, 128, 133, 138, 147, 164, 173
bilingualism
  additive, 104, 108, 164–166, 170, 173
  emerging, 108
  policy, 140
  subtractive, 109, 165–166
  unequal, 109
biological species, 4, 21, 117, 128
biological time, 3
bit-string model
  Kosmidis, Halley, and Argyrakis, 119
  Schultze and Stauffer, 117, 119

carrying capacity, 118, 121, 149, 151, 155, 160
change rate, 13
choreme, 48, 50, 52, 72
clade, 20, 34–35, 37–40

cladistics, 27, 37, 59, 64
cluster, 10, 13, 20, 29, 35, 37, 56, 77, 112, 175
coalescence, 16
coarse-graining, 100–101
cognitive system, 3, 106
cohesiveness, 80–81, 86
communication failure, 113
communication strategy, 111
communication success, 113
  active, 106
  passive, 106
competition, 1, 11, 97, 118, 128, 137, 143, 168,
complex network, 1, 21, 23–24, 26, 115–116, 161
complex system, 1, 21, 100, 161, 177
complex systems theory, 5–6, 177
complexity, 1, 12, 99 116, 129, 163, 177
complexity theory
  bit, 1, 9
  communal, 1
  constitutional, 1
  interactional, 1
  structural, 5
connectivity, 23, 26
consensus, 110, 112–115, 165
consonant, 17, 19, 90
coronal, 20
creole, 3, 13, 29–31, 103, 144
cross-fertilization , 29

Darwinism, 28–30
database
　ALTO, 67
　ASJP, 19–20
　EAS, 57, 164
degree, 21, 23–27, 78, 100, 108, 131
degree distribution
　average, 23
deletion, 17, 61
demographic process, 101
dendrogram, 55, 91
dialect, 2, 9, 17, 28, 37, 47, 54, 59, 61, 67, 73, 86, 102, 124, 164
dialect continuum, 9, 68, 79, 102
dialect
　attributive, 40
　blind, 40
dialectometry, 37, 64
diffusion
　cultural, 16, 151
diglossia, 5, 104–105, 164
dilalia, 104, 105
diphthongization, 18
dispersal
　inhomogeneous, 153, 157
dissimilation, 16–18
distance matrix
　Hamming, 42, 44
　Jaro, 44
　longest common subsequence, 44
　metric, 41–42, 56–57
diversity, 4, 12, 68, 97, 105, 121,
dorsal, 20

ecological factors , 100, 129, 142
ecological process, 4
ecology, 4–6, 21, 101, 161
edge, 21, 26, 54
edit distance , 17, 41, 44
edit operations, 41–44
equilibrium point , 131–132, 137, 140
Erdar, 109, 164, 168

ergative alignment, 3, 68
Euskalkia, 104, 109
evolution, 1, 11, 110, 118, 147, 164, 178
evolutionary linguistics, 13, 30–31, 37
exaptation, 11–14
exaptive system, 14
expanding normalization, 108
extinction, 29, 35, 98, 108, 152, 163

feed, 17, 59
fieldwork, 12
fitness
　social linguistic, 121
forgetting process, 120
fortition, 16–17
frontness, 20

gender variation, 170, 172
gene transfer, 10
genealogical ramification, 29, 34
genera, 14–15, 19–20, 28, 35, 37
genus, 14, 20, 32, 38–40
geographical barrier, 142, 157
geography, 39, 77, 83, 90, 117, 151
geolinguistics, 31
globally stable, 139–140
glottometric diagram, 81, 86, 93
grammatical calquing, 14
grammatical function , 18
grammatical robustness, 28
graph
　complete, 24, 76
　Erdős-Rényi random, 25
　random, 25–26

heat map, 88
heterogeneous, 19, 26, 32, 56, 142, 156
historical glottometry , 10, 23, 80, 85, 93
Holling factor, 162–163
homogeneous, 21, 25, 39, 125, 142, 157
hub, 23

incipient activism, 108
incipient mobilization, 108
incipient normalization, 108
index
  agglutination, 32
  compounding, 33
  derivation, 33
  inflection, 33
  prefixation, 33
  similarity, 44–45, 68, 71–75
  suffixation, 33
  synthesis, 32
innovation
  conflicting, 80
  exclusively shared, 80–81, 86
  irregular, 85–86
  regular, 85–86
insertion, 17, 44
interaction, 3, 22, 113, 120, 163
interdisciplinary, 1, 6, 97, 130
IPA, 9, 10, 84
isogloss, 39, 80, 86–87

language acquisition, 98, 112, 174
language change, 2, 13, 97, 110, 153, 177
language cognition, 2
language comparison, 9, 10, 43
language competition, 2, 4–5, 101, 128, 168
language contact, 13
language death, 13, 106–107
language diffusion
language dynamics, 1, 13, 97, 100, 110, 160, 164, 178
language ecology, 5
language evolution, 2, 11, 79, 102, 116, 164
language faculty, 11, 103,
language family, 10, 13, 28, 79
language maintenance, 129
language origin, 12, 178
language planning, 5, 164, 172
language shift, 78, 107, 147, 154, 166
language similarity

language use
  dominant, 145, 148, 164
  dominated, 145
  donor, 30
  endangered, 4, 46, 98, 106, 134, 148
  minority, 5, 105, 129, 141, 146–148, 165
  national, 105, 174
  natural, 16, 18, 102
  proto-, 10–11, 35, 79, 84
  vehicular, 98, 105,
lateral, 20, 52, 141
lattice constant, 21
learning process, 10
lenition, 16, 18, 32
Levenshtein distance
  Damerau-, 44
  non-weighted average, 59
  relative, 43
  weighted average, 59
Levenshtein matrix, 43, 88
lexical variation, 74, 76
lexicology, 67
lineage, 12
lingua franca, 37, 98, 102, 109
linguistic area, 10
linguistic barrier, 153, 157
linguistic change, 29, 99
linguistic community, 6, 108, 128
linguistic policy, 107, 168
linguistic stock, 14–15, 19, 28, 32, 34
linguistics
  comparative, 28–31, 68
  computational, 3, 98
  general, 29
link, 21–23, 54, 64, 72, 89, 148
linkage, 9, 79
local endemicity, 108
local resilience, 108
logistic dynamics, 150
logistic term, 149, 151
loop, 12, 17, 59, 120, 122
Lotka-Volterra equation, 101

Malthus rate, 118, 149–150, 160, 162
Malthusian principle, 29
Mayan language, 35, 37, 66
Mazatec, 4, 13, 32, 45–47, 124
mean field equation, 137, 143
meme, 2
metathesis, 17
minimum spanning tree, 54, 92,
model
   AB, 137–138, 144, 152
   Abrams and Strogatz, 129–130
   Baggs and Freedman, 161
   Cedergren's, 17–18
   continuous, 6, 149
   conversation, 147
   evolutionary, 97
   family tree, 79
   individual-based, 22, 100–101, 117, 137, 143
   language dynamics, 97, 100–101, 110, 163, 178
   macroscopic, 101, 130
   mean field, 101
   mesoscopic, 101
   Minett and Wang, 134, 141
   naming game, 112, 143
   natural selection, 97
   Nowak, 102, 112, 116
   Pinasco and Romanelli, 150
   semiotic dynamics, 97, 101, 110
   SFVL, 59, 61
   socio-cognitive, 2
   three-state, 128, 133–134, 138, 149, 161
   two-state, 128–130, 149, 151, 161
   Viviane De Oliveira, 117, 121
   voter, 100, 137, 152
monolingual, 101, 133, 135, 143, 161, 168
morphology, 32, 67–68, 116
multidimensional scaling, 10, 56, 92
multidimensional space, 56
multilingual, 1, 15, 105, 147, 164
multilingualism, 5, 105, 129, 141, 173
mutation, 11, 101, 118, 126

mutual intelligibility, 15, 22, 46, 102, 125

naming game, 112, 143
Native Land website, 82
natural selection, 11, 97, 128
nearest neighbor map, 90
Neogrammarian, 15, 29, 31
Neolinguistica, 29
network, 1, 17, 21, 28, 54, 61, 63, 82, 92, 101, 141, 151, 168, 173
network of interactions, 22
network of linguistic similarity, 22
network of mutual intelligibility, 22
network topology
   connected, 24, 48, 56, 72, 138,
   directed, 23
   fully connected, 24, 48, 56, 72, 138, 151, 161
   regular, 21, 25
   scale-free, 26–27
   semantic, 22
   small-world, 23, 25
   undirected, 23
niche, 13, 177
node, 21, 54, 72
nonlocal dispersion, 157
noun, 18, 34, 52,
numerical experiment, 99, 111, 125, 157
numerical simulation, 4, 10, 99, 125, 159
Numic, 6, 17, 82, 85, 93
Numic homeland, 83
Numic spread hypothesis, 82, 84, 93

open taxonomy, 34
opinion dynamics, 2, 13

path length
   average, 23, 26–27
   mean, 25
pattern, 2, 17, 39, 61, 79, 105, 164, 175
pause, 18–19
phase portrait, 132, 138, 139

phonation, 20
phonology, 11, 32, 67, 73, 77, 116
phonotactic, 18
phyla, 15, 28
phylogenetic, 14, 37, 47, 64, 103
phylogenetic isolate, 14
phylogram, 34, 38, 40,
Pidgin, 13, 30–31, 103
Poissonian distribution, 25
population dynamics, 102, 111, 149, 160
position, 18, 35, 47, 90, 152, 159
power law, 23, 26–27, 115, 117, 138
prestige, 15, 28, 77, 108, 129, 134

reaction–diffusion equation, 152, 153, 158
reactivated proficiency, 106
regional variation, 57, 168
register, 13, 15, 19, 104
regular lattice, 21, 24, 133
repertoire, 3, 104–106, 172
reproduction, 111, 118, 120
residual colloquiality, 108
revitalization, 106
rhoticity, 20

segment, 17, 19, 40, 108
selection, 2, 13, 97, 128, 157
self-organization, 3, 30, 40, 103
semiotic dynamics, 97, 101, 110, 112, 144
sibilant, 18–19
social time, 3
sociolinguistic, 5, 19, 103, 107, 146, 164
sociolinguistics, 5, 103, 149
sonority, 20
sound change, 13, 16, 38, 116
sound correspondence, 19–20
spanning tree, 28, 54, 92,
speciation
   allopatric, 10
   parapatric, 10
speech style, 18, 77
sprachbund, 10

spreading, 2, 22, 40, 64, 126, 161
Stammbaum, 28, 56
Stammbaumtheorie, 79
statistical mechanics, 1, 4, 99
strengthening, 16–17, 107
string, 10, 41, 72, 119, 124
string metric, 10, 41
structural embedding, 18
structure, 6, 59,
subgraph, 24, 76, 78
subgroupiness, 81
substitution, 44, 123
syllable, 17–19,
syntactic category, 18

taxonomy, 20, 29, 34, 57
threshold, 47, 54, 72, 116, 123,
time evolution, 98, 118, 120–121, 147
token, 13, 28, 35
trait
   converging, 29
   cultural, 2, 16
transition rate, 131, 135, 143
transmission
   horizontal, 134, 162
   vertical, 118, 134, 162
transposition, 44
tree, 10, 27–28, 54, 79, 92
Tseltal, 6, 17, 35, 66–68, 73, 76
Turing pattern, 156
typology, 19, 31–32, 34, 68, 77

universal grammar, 3, 38, 68, 103

variation, 12, 19, 57, 68, 74, 168, 170
vehicular, 98, 109, 173
vernacular, 104, 108, 164, 173
vertice, 21, 28, 72
vocal tract, 12
void, 59
volatility
   parameter, 130–131, 148

high, 131
low, 131
neutral, 132
vowel, 17–19, 38, 61

WALS, 15
wave front, 160
wave theory, 10, 79–80
weakening, 16